Editor
Sara Connolly

Managing Editor
Ina Massler Levin, M.A.

Illustrator
Clint McKnight

Cover Artist
Brenda Di Antonis

Art Production Manager
Kevin Barnes

Art Coordinator
Renée Christine Yates

Imaging
Ralph Olmedo, Jr.

Publisher
Mary D. Smith, M.S. Ed.

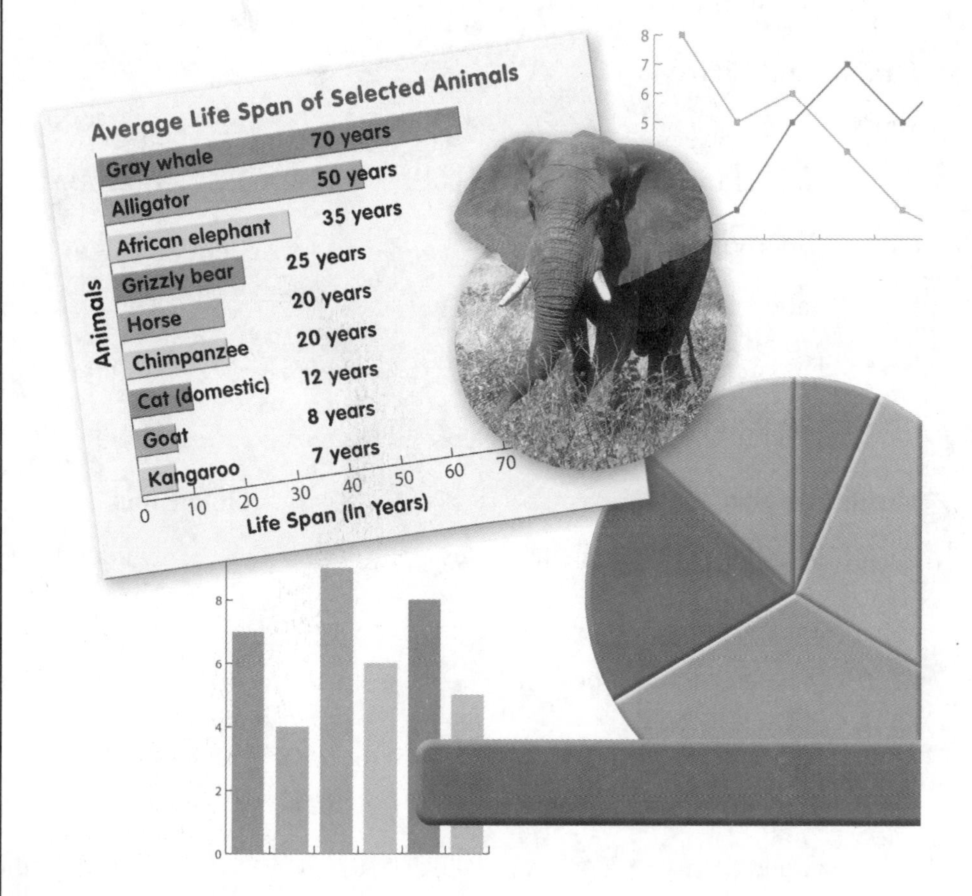

Author

Shelle Russell

Teacher Created Resources, Inc.
6421 Industry Way
Westminster, CA 92683
www.teachercreated.com
ISBN-1-4206-8016-1

Table of Contents

Introduction

The World Almanac for Kids® Charts and Graphs combines the wealth of information of *The World Almanac for Kids* with the skills of creating graphs and charts, which are part of the national standards required for children in the third and fourth grade. Children become fact wizards as they gather information, then create different kinds of charts and graphs. These charts and graphs provide children with hours of stimulating mental activities as well as opportunities to fine-tune learning.

The activities in this book are designed to help develop the following:

- graphing skills
- deciphering/locating data
- building knowledge of the world and local events

How to Use *The World Almanac for Kids Charts and Graphs*

The World Almanac for Kids Charts and Graphs is divided into sections by subject: bar graphs, circle graphs, line graphs, line plots, picture graphs, stem and leaf plots, time lines, and coordinate graphs.

- **Introductory Pages**
  The introductory pages provide samples of each type of graph or chart for both the teacher and the students. These can be reproduced for the students to use as references, used in centers or bulletin boards, or made into overheads for use as sample lessons.

- **Information Pages**
  Information pages come directly from *The World Almanac for Kids* in a variety of subject areas. Children will need to sift through the information on these pages to create their charts and graphs. Encourage students to mark or highlight information on those pages to assist them in gathering data and using the information correctly.

- **Charting and Graphing Pages**
  Charting and graphing pages require children to create their graphs and then use the graphs to answer a series of questions. Some activities require children to complete charts before making their graphs. Each page also includes a challenge activity to extend the graphing lesson.

  Note to the teacher: You may need to adjust the difficulty level of these activities by having peer tutors assist the students who are having difficulty locating the information. Encourage students to use test-taking strategies such as highlighting, circling, and underlining on information pages to assist them. With these strategies, even struggling students can become successful at identifying key information. As they continue practicing, they will become comfortable with many different types of graphs.

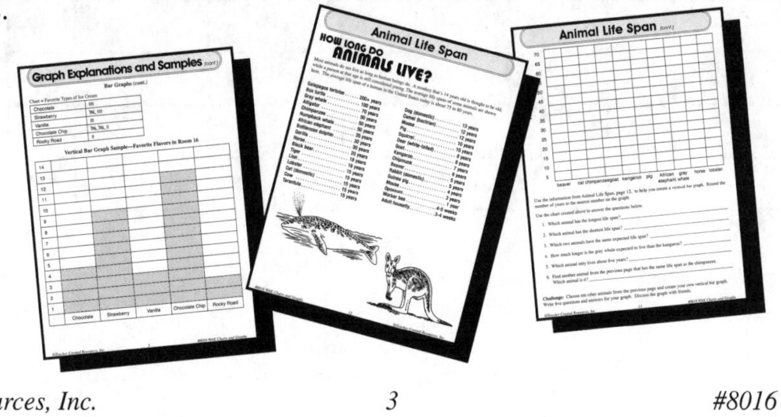

About This Book

This book incorporates real life information from our world into a variety of graphing activities. The entire book is divided into eight different sections. For each activity, information is provided directly from *The World Almanac for Kids 2006.* Sometimes charts are also used to collect data before the actual graphing occurs. Graphs follow the information and display collected data. Step-by-step instructions are included to facilitate the information and the type of graphing that will be completed. All graphing sections can be extended for more graphing activities. The activities are presented in order of difficulty, so the teacher can use the activity which best suits the level desired. When the graphs are completed, they can be used as assessments, centers, bulletin boards, or skill-building activities. This book is a useful tool for reinforcing and enhancing the art and mastery of a variety of forms of graphing.

Graph Explanations and Samples

Bar Graphs

Bar graphs are graphs that use either horizontal or vertical bars to compare information from several categories. Before the data is entered on the graph, it is always wise to collect information on a chart. Below is a sample of a horizontal bar graph. A vertical bar graph sample can be found on page 5.

Informational Chart = Favorite Types of Ice Cream

Chocolate	IIII
Vanilla	III
Strawberry	~~IIII~~ IIII
Chocolate Chip	~~IIII~~ ~~IIII~~ II
Rocky Road	II

Horizontal Bar Graph Sample—Favorite Flavors in Room 16

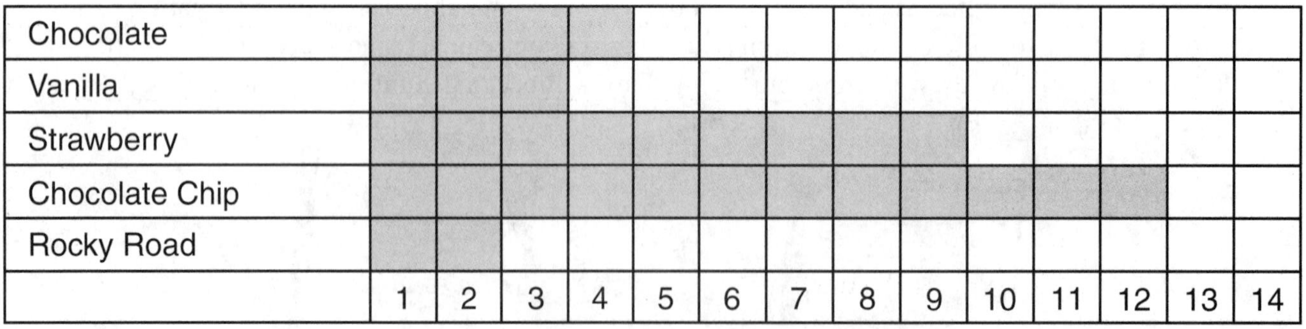

Chocolate														
Vanilla														
Strawberry														
Chocolate Chip														
Rocky Road														
	1	2	3	4	5	6	7	8	9	10	11	12	13	14

Graph Explanations and Samples *(cont.)*

Bar Graphs *(cont.)*

Chart = Favorite Types of Ice Cream

Chocolate	IIII
Strawberry	~~IIII~~ IIII
Vanilla	III
Chocolate Chip	~~IIII~~ ~~IIII~~ II
Rocky Road	II

Vertical Bar Graph Sample—Favorite Flavors in Room 16

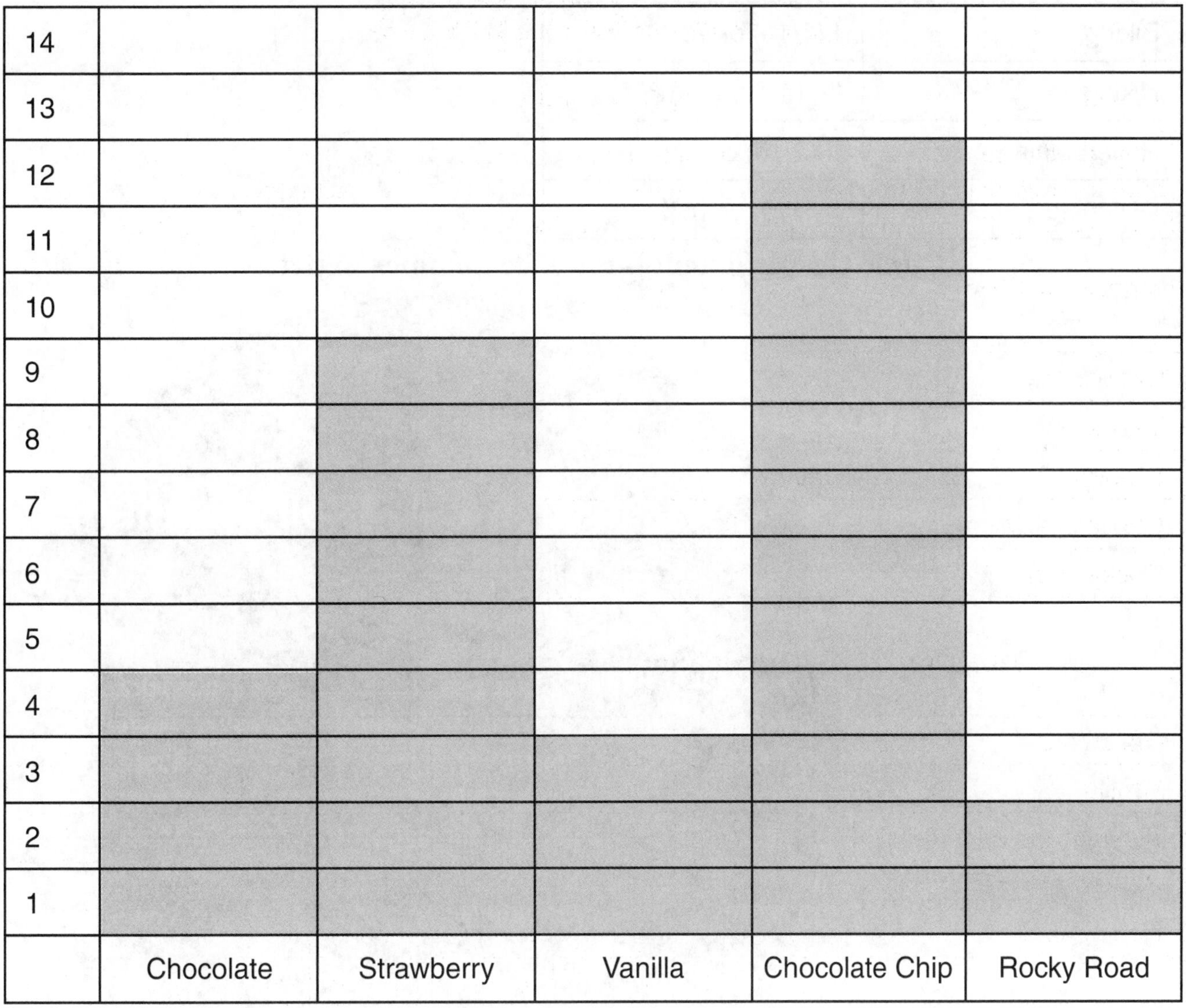

14					
13					
12					
11					
10					
9					
8					
7					
6					
5					
4					
3					
2					
1					
	Chocolate	Strawberry	Vanilla	Chocolate Chip	Rocky Road

Graph Explanations and Samples *(cont.)*

Circle Graphs

Another name for a circle graph is a pie chart. Data is used to compare parts of a whole. In other words, a pie chart shows the relationship between a whole entire circle to the variety of pieces or percentage into which it is divided. Below is an example of a circle graph. To find percentages for the circle graph, create fractions. Place the number collected on top (numerator) and the total number of people or things polled on the bottom (denominator). To find percentages for the circle graph, multiply the numerator by 100 and divide that answer by the denominator.

Collecting data: Student interviewed 10 people about preference of five summer activities.

Informational chart

Swimming	II (2/10) or 20%
Biking	I (1/10) or 10%
Hiking	IIII (4/10) or 40%
Rollerblading	II (2/10) or 20%
Watching TV	I (1/10) or 10%

Circle Graph Sample—Favorite Summer Activites

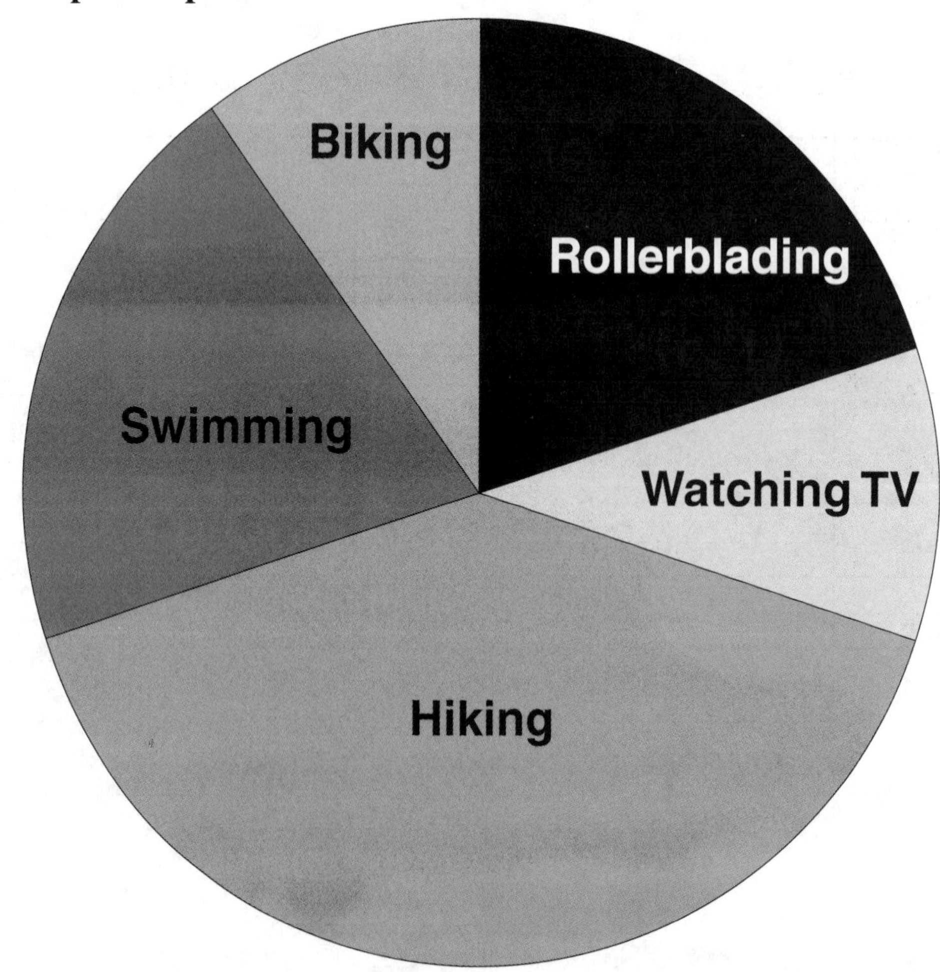

Graph Explanations and Samples *(cont.)*

Line Graphs

A line graph is composed of one continuous line where data is joined together. It shows how things change over time. Line graphs may include single data which is collected over a period of time, or data that includes more than one category. Multiple Line graphs include data from several different places or things. Use several different colors to indicate the different data.

Chart information:

Weather in Bemidji, Minnesota	Highs (°F)	Lows (°F)
June 1	76	72
June 2	76	72
June 3	74	63
June 4	80	75
June 5	76	70
June 6	74	68

Single Line Graph Sample—Graphing the High Temperatures in Bemidji

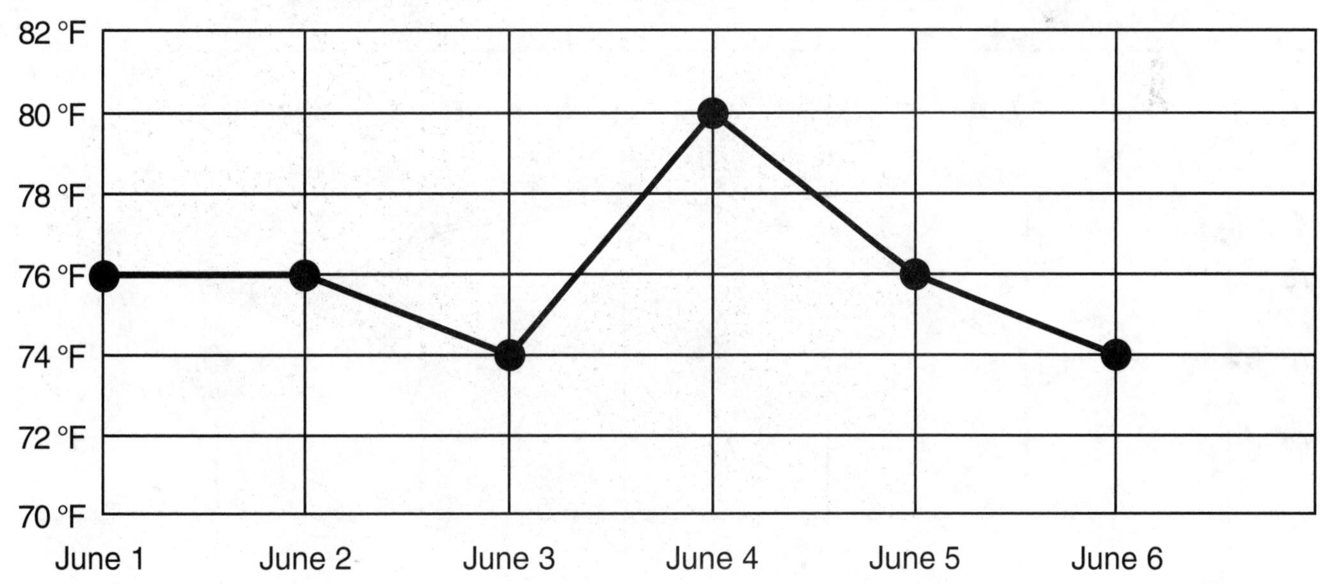

Graph Explanations and Samples *(cont.)*

Line Graphs *(cont.)*

Chart information:

Weather in St. Paul, Minnesota	Highs (°F)	Lows (°F)
June 1	79	76
June 2	75	72
June 3	69	64
June 4	79	75
June 5	70	69
June 6	76	62

Weather in Bemidji, Minnesota	Highs (°F)	Lows (°F)
June 1	76	72
June 2	76	72
June 3	79	63
June 4	80	80
June 5	79	76
June 6	75	70

Multiple Line Graph Sample—Graphing the High Temperatures in Bemidji and St. Paul

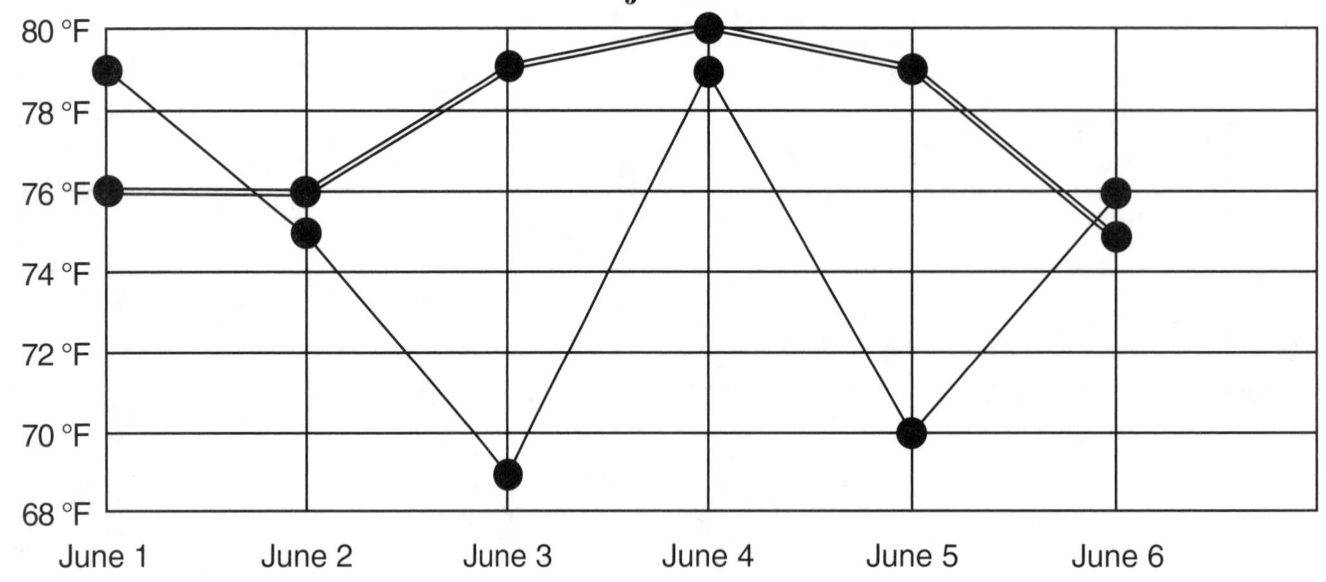

Single line = St. Paul
Double line = Bemidji

Graph Explanations and Samples *(cont.)*

Line Plots

Line plots give you a clear picture of recorded data. Mode, range, and median can be determined easily with line plots. Numerical data is recorded by using Xs above a number line. One X is made to represent information collected. X may equal 1, or X may represent 10 or more, depending on the graph.

Informational chart: Number of Buttons on Clothes

Sally	6
Sammy	3
Frankie	2
Joe	9
April	6
Jennifer	3
Jessica	5
Amy	7
Fred	7
Karen	2

X=1

Line Plot Sample—Number of Buttons on Clothes

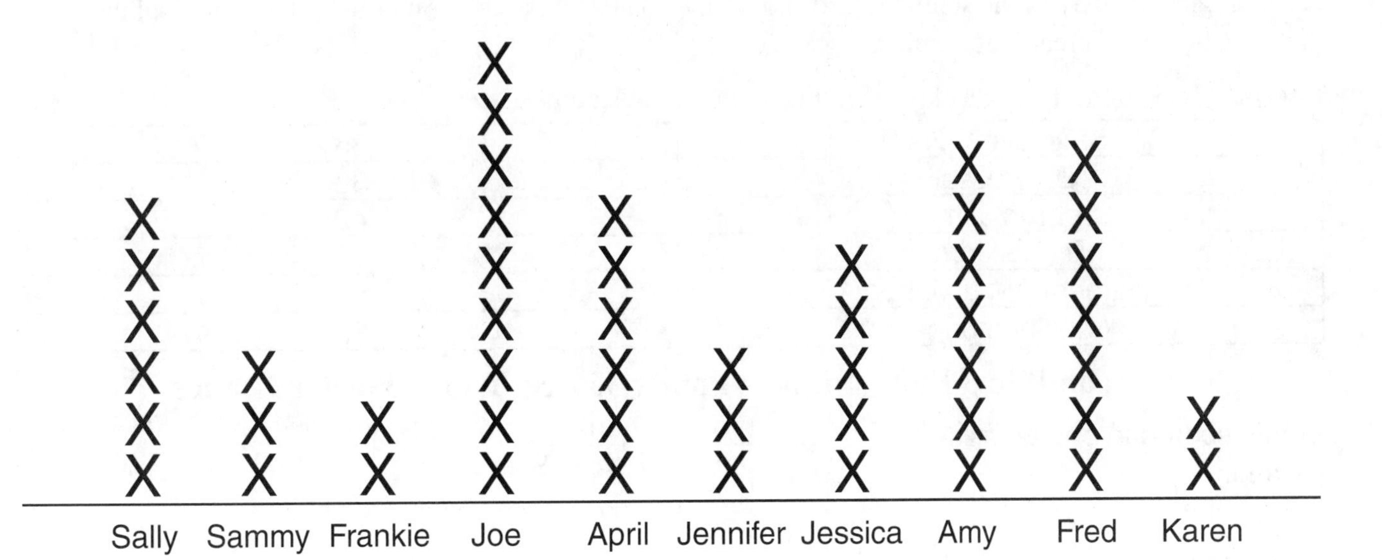

Picture Graphs

Picture graphs are also known as pictographs. They are made up of pictures, simple diagrams, or illustrations that symbolize real things from society.

Informational chart: Books read over one week in this family

Shelle	2
Cindy	2
Kerry	0
Christine	4
Val	1
Tim	5
Nate	3

Picture Graph Sample—Books Read in One Week

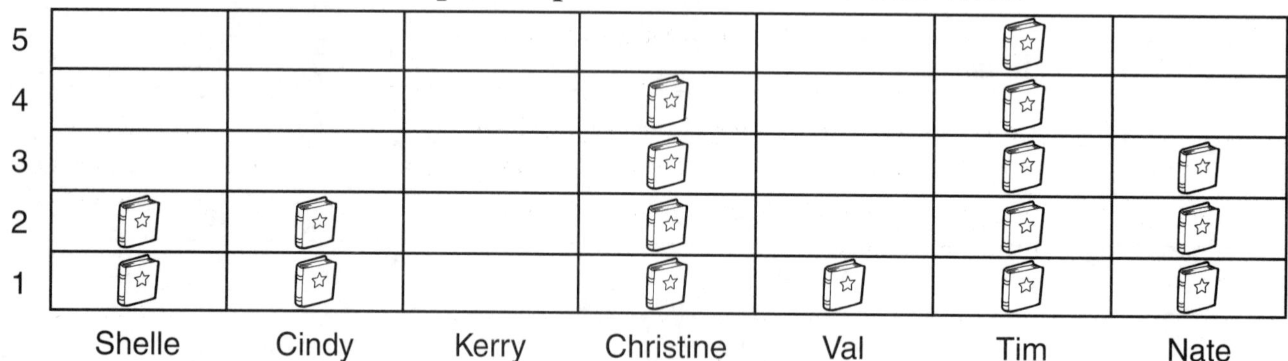

Stem and Leaf Plots

Numerical information is organized in such a way that the numbers themselves make an organized, chronological display in the stem and leaf plot graph. The stem and leaf plot uses the numerical data that is collected to create the display.

Informational chart: Famous players points in one week's games

Shaquille O'Neil	35
Bill Russell	39
Lebron James	42
Michael Jordan	65
Kareem Abdul-Jabbar	51
Wilt Chamberlain	39

Stem and Leaf Plot Sample—Points Scored in One Week's Games

Points scored in one week's games:

Stem	Leaf
3	5 9 9
4	2
5	1
6	5

Key: 5 1 = 51 points

Graph Explanations and Samples *(cont.)*

Time Lines

Time lines display information on a number line. A time line can show information for years, days, hours, or minutes. Events are plotted in the order in which they occur. Time lines can be made for the past or for the future depending on the information.

Time Line Sample—Mrs. Winkler's Life

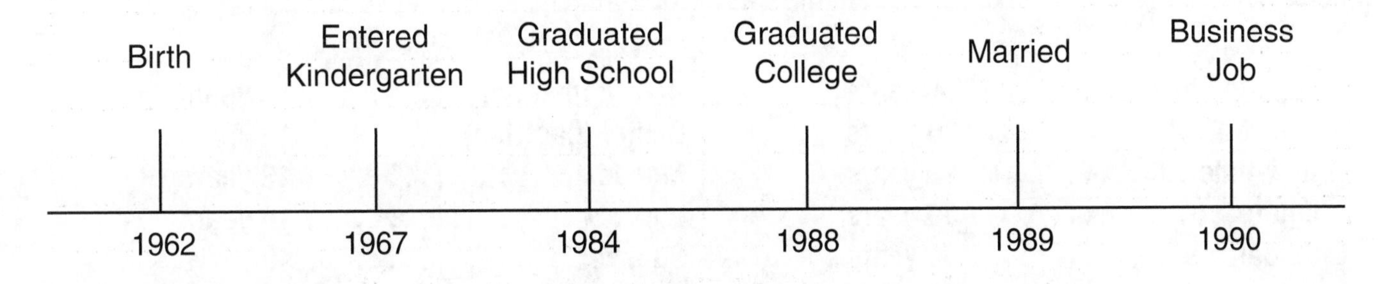

Coordinate Graphs

Coordinate graphing can be compared to finding addresses in a neighborhood. When locating information using coordinates, the two numbers in the ordered pair always remain the same. The first number is always the horizontal coordinate; the second number is always the vertical coordinate. Coordinate graphing can be used for any subject area. It is similar to the game Battleship™, in which you sink ships by calling coordinates.

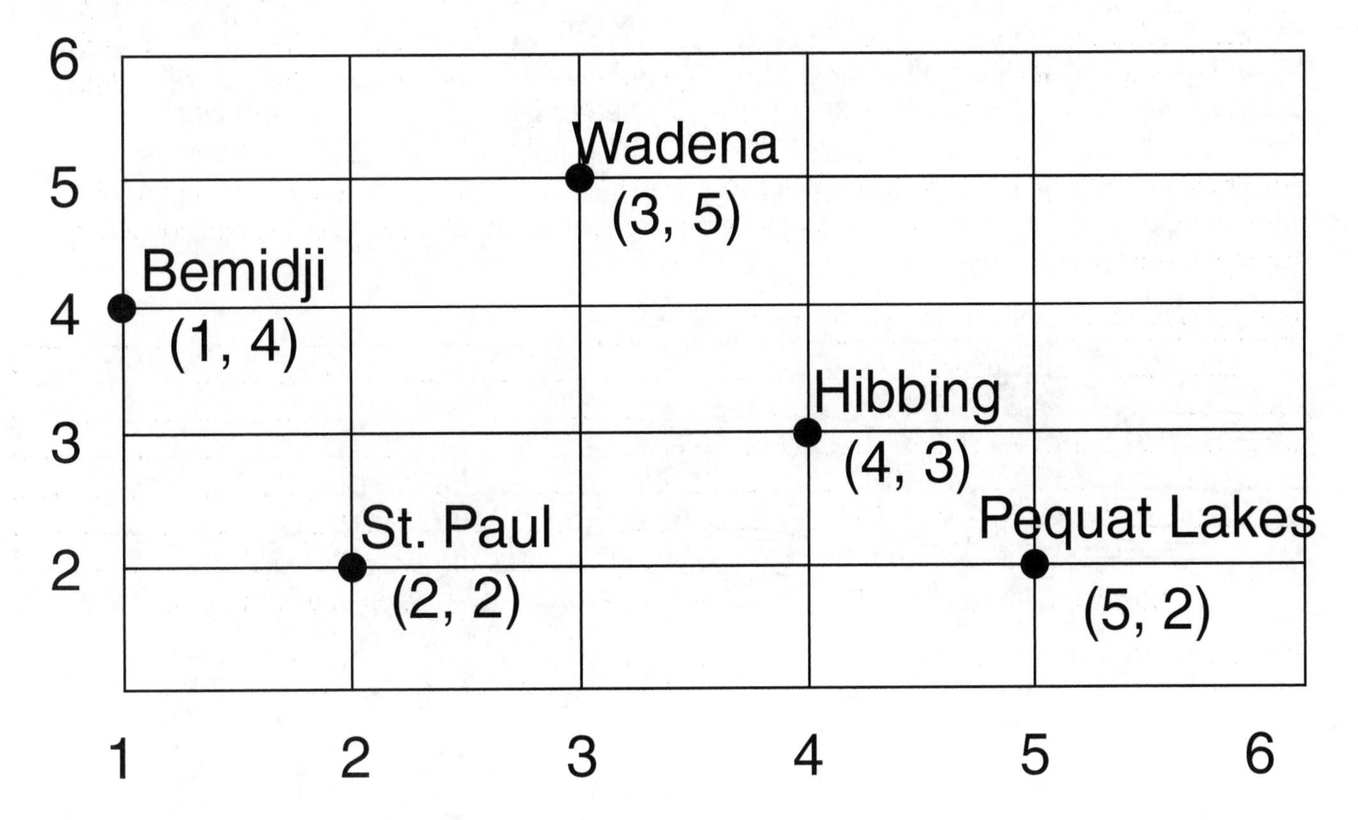

Animal Life Span

HOW LONG DO ANIMALS LIVE?

Most animals do not live as long as human beings do. A monkey that's 14 years old is thought to be old, while a person at that age is still considered young. The average life spans of some animals are shown here. The average life span of a human in the United States today is about 75 to 80 years.

Galapagos tortoise . . . . . . 200+ years
Box turtle . . . . 100 years
Gray whale . . . . 70 years
Alligator . . . . 50 years
Chimpanzee . . . . 50 years
Humpback whale . . . . 50 years
African elephant . . . . 35 years
Bottlenose dolphin . . . . 30 years
Gorilla . . . . 30 years
Horse . . . . 20 years
Black bear . . . . 18 years
Tiger . . . . 16 years
Lion . . . . 15 years
Lobster . . . . 15 years
Cat (domestic) . . . . 15 years
Cow . . . . 15 years
Tarantula . . . . 15 years
Dog (domestic) . . . . 13 years
Camel (bactrian) . . . . 12 years
Moose . . . . 12 years
Pig . . . . 10 years
Squirrel . . . . 10 years
Deer (white-tailed) . . . . 8 years
Goat . . . . 8 years
Kangaroo . . . . 7 years
Chipmunk . . . . 6 years
Beaver . . . . 5 years
Rabbit (domestic) . . . . 5 years
Guinea pig . . . . 4 years
Mouse . . . . 3 years
Opossum . . . . 1 year
Worker bee . . . . 4-5 weeks
Adult housefly . . . . 3-4 weeks

Animal Life Span *(cont.)*

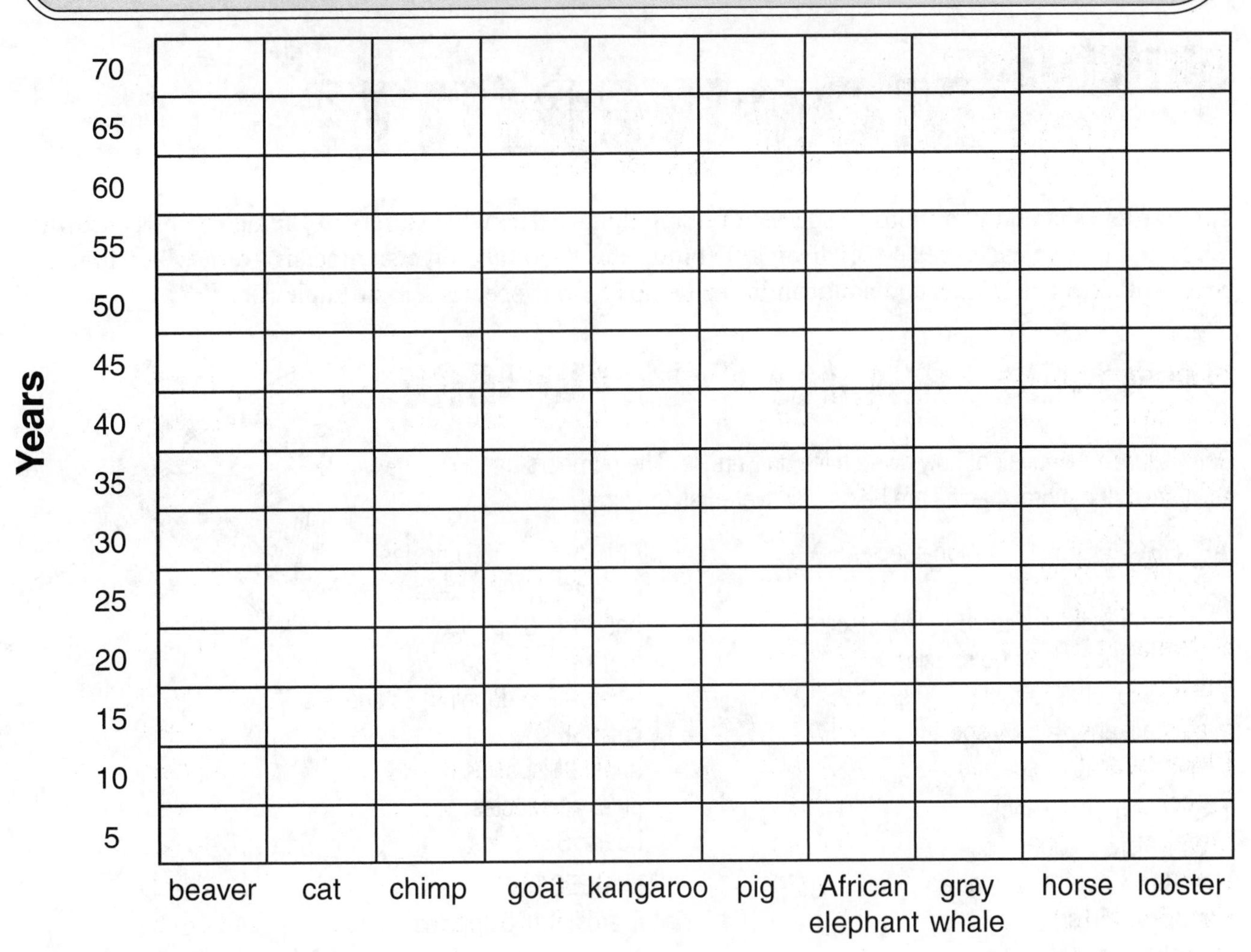

Use the information from Animal Life Span, page 12, to help you create a vertical bar graph. Round the number of years to the nearest number on the graph.

Use the chart created above to answer the questions below.

1. Which animal has the longest life span? ____________________
2. Which animal has the shortest life span? ____________________
3. Which two animals have the same expected life span? ____________________
4. How much longer is the gray whale expected to live than the kangaroo? ____________________
5. Which animal only lives about five years? ____________________
6. Find another animal from the previous page that has the same life span as the chimpanzee. Which animal is it? ____________________

Challenge: Choose ten other animals from the previous page and create your own vertical bar graph. Write five questions and answers for your graph. Discuss the graph with friends.

Biodiversity in Species

WHAT IS BIODIVERSITY?

The Earth is shared by millions of species of living things. The wide variety of life on Earth, as shown by the many species, is called "biodiversity" (bio means "life" and diversity means "variety"). Human beings of all colors, races, and nationalities make up just one species, Homo sapiens.

SPECIES, SPECIES EVERYWHERE

Here is just a sampling of how diverse life on Earth is. The numbers are only estimates, and more species are being discovered all the time!

ARTHROPODS (1.1 million species)
- insects: 750,000 species
  - moths & butterflies: 165,000 species
  - flies: about 122,000 species
  - cockroaches: about 4,000 species
- crustaceans: 44,000 species
- spiders: 35,000 species

FISH (24,500 species)
- bony fish: 23,000 species
- skates & rays: 450 species
- sharks: 350 species
- seahorses: 32 species

BIRDS (9,000 species)
- perching birds: 5,200-5,500 species
- parrots: 353 species
- pigeons: 309 species
- raptors (eagles, hawks, etc.): 307 species
- penguins: 17 species
- ostrich: 1 species

MAMMALS (9,000 species)
- rodents: 1,700 species
- bats: 1,000 species
- monkeys: 242 species
- whales and dophins: 83 species
- cats: 38 species
- apes: 21 species
- pigs: 14 species
- bears: 8 species

REPTILES (8,000 species)
- lizards: 4,500 species
- snakes: 2,900 species
- tortoises & turtles: about 294 species
- crocodiles & alligators: 23 species

AMPHIBIANS (5,000 species)
- frogs & toads: 4,500 species
- newts & salamanders: 470 species

PLANTS (260,000 species)
- flowering plants: 250,000 species
- bamboo: about 1,000 species
- evergreens: 550 species

Biodiversity in Species *(cont.)*

Species	monkeys	pigeons	parrots	sharks	salamanders
450					
400					
350					
300					
250					
200					
150					
100					
50					

Use the information from Biodiversity in Species, page 13, to complete the vertical bar graph. Round your answers to the nearest number on the graph. Use only the animals assigned on the graph.

1. Which creature listed above has the most species? ____________________

2. Which has the least amount of species?____________________

3. How many more species of salamanders are there than parrots?____________________

4. How many more species of salamanders are there than monkeys? ____________________

5. How many more species of salamanders are there than sharks? ____________________

Challenge: Create a horizontal bar graph using 10 other species from the previous page. Create questions and answers to go with your bar graph.

Tallest Buildings in the World

TAIPEI 101,
Taipei, Taiwan (2004)
Height: 101 stories,
1,671 feet

PETRONAS TOWERS 1 & 2, Kuala Lumpur, Malaysia (1998) Height: each building is 88 stories, 1,483 feet

SEARS TOWER,
Chicago, Illinois (1974)
Height: 110 stories,
1,450 feet

JIN MAO TOWER,
Shanghai, China (1998)
Height: 88 stories,
1,380 feet

TWO INTERNATIONAL FINANCE CENTRE,
Hong Kong, China (2003) Height: 88 stories,
1,362 feet

CITIC PLAZA,
Guangzhou, China (1997)
Height: 80 stories,
1,283 feet

Tallest Buildings in the World *(cont.)*

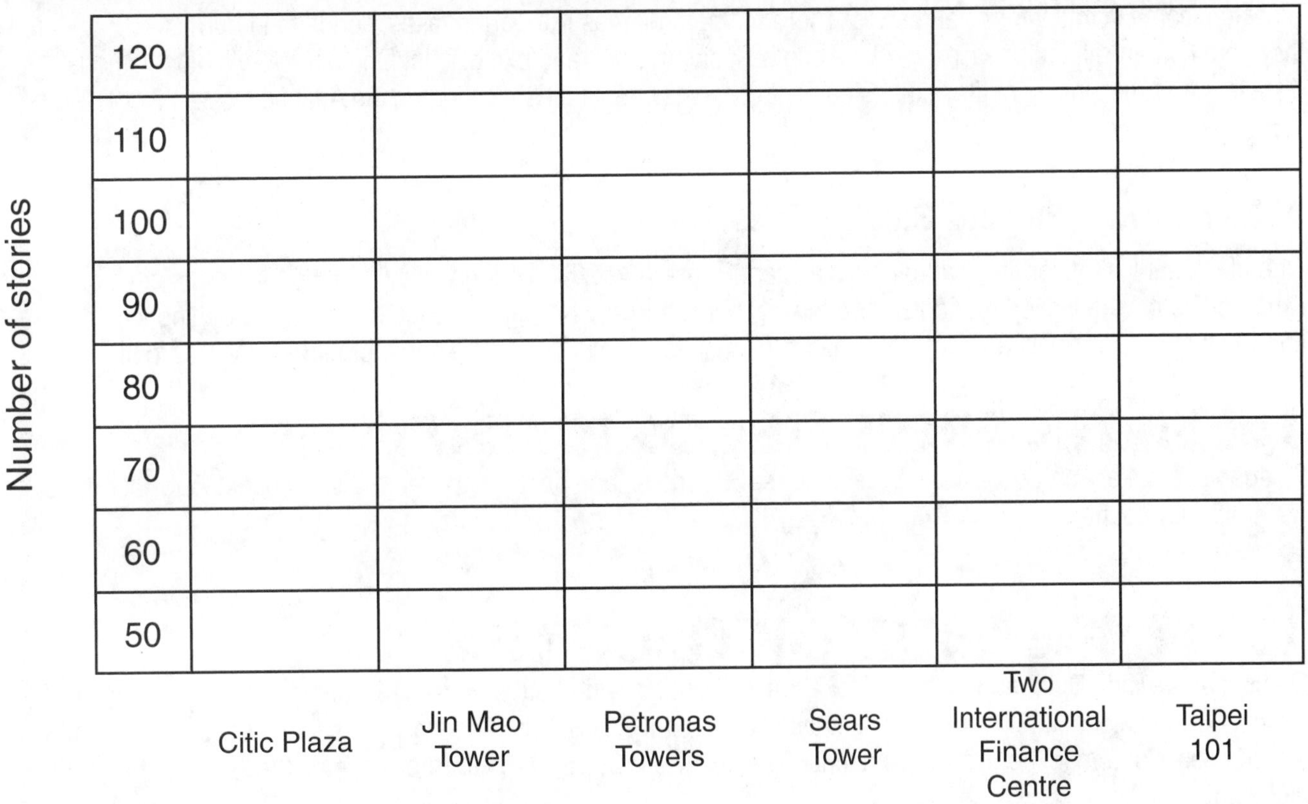

Number of stories

	Citic Plaza	Jin Mao Tower	Petronas Towers	Sears Tower	Two International Finance Centre	Taipei 101
120						
110						
100						
90						
80						
70						
60						
50						

Use the information from Tallest Buildings in the World, page 16, to complete the vertical bar graph. Record only the buildings named on the graph. Round your answers to the nearest ten.

Use the graph to answer the questions below.

1. Which tower has the most stories? ______________________

2. Which towers are the same number of stories?______________________

3. How many more stories are in the Taipei than in the Citic?______________________

4. How many more stories are in the Jin Mao than the Petronas? ______________________

5. If you could stack the Petronas, the Sears, and the Citic on top of each other, how many stories would you have? ______________________

Challenge: Create a horizontal bar graph using the information regarding the height of the buildings as recorded in feet from page 17. Create five questions and answers to go with your bar graph.

Disasters

Disasters come in many forms and sizes. There are natural ones like earthquakes, floods, and hurricanes. And there are human-related ones like airplane crashes, shipwrecks, and explosions. One of history's most famous disasters was the sinking of the luxury steamship *Titanic* (when it hit an iceberg in the Atlantic Ocean) in 1912; 1,503 people died.

ALL ABOUT . . . Famous Big City Fires

One of the earliest major fires happened in ancient Rome in A.D. 64. Legend has it that the blaze was set by the Emperor Nero, who played his fiddle while the city burned. Many historians doubt this story. (One thing they do know is that he didn't play a violin, which hadn't been invented yet.) Here are some other big city fires in history:

1666: THE GREAT FIRE OF LONDON

It began in a bakery on September 2, lasted for several days, and destroyed over 13,000 houses. Only four dead were officially counted.

1871: THE GREAT CHICAGO FIRE

Coming on a windy October 8 at the end of a hot, dry spell, it wiped out an area of 3 square miles in the center of the city, killing at least 300 people and destroying over 18,000 buildings. The story goes that it began when a cow kicked over a lantern in the barn of Catherine O'Leary. When Chicago rebuilt, they called it the "The Second City," a nice name that lives today.

1904: BALTIMORE FIRE

This fire started February 7. Eighty-six blocks of the business district were affected, 1,500 buildings burned, and 2,500 businesses were damaged or destroyed. Remarkably, only four people died, all firefighters.

1906: SAN FRANCISCO EARTHQUAKE AND FIRE

A massive earthquake early on the morning of April 18 set off a series of destructive fires. The death toll was over 3,000, and damage was estimated at $500 million.

Disasters *(cont.)*

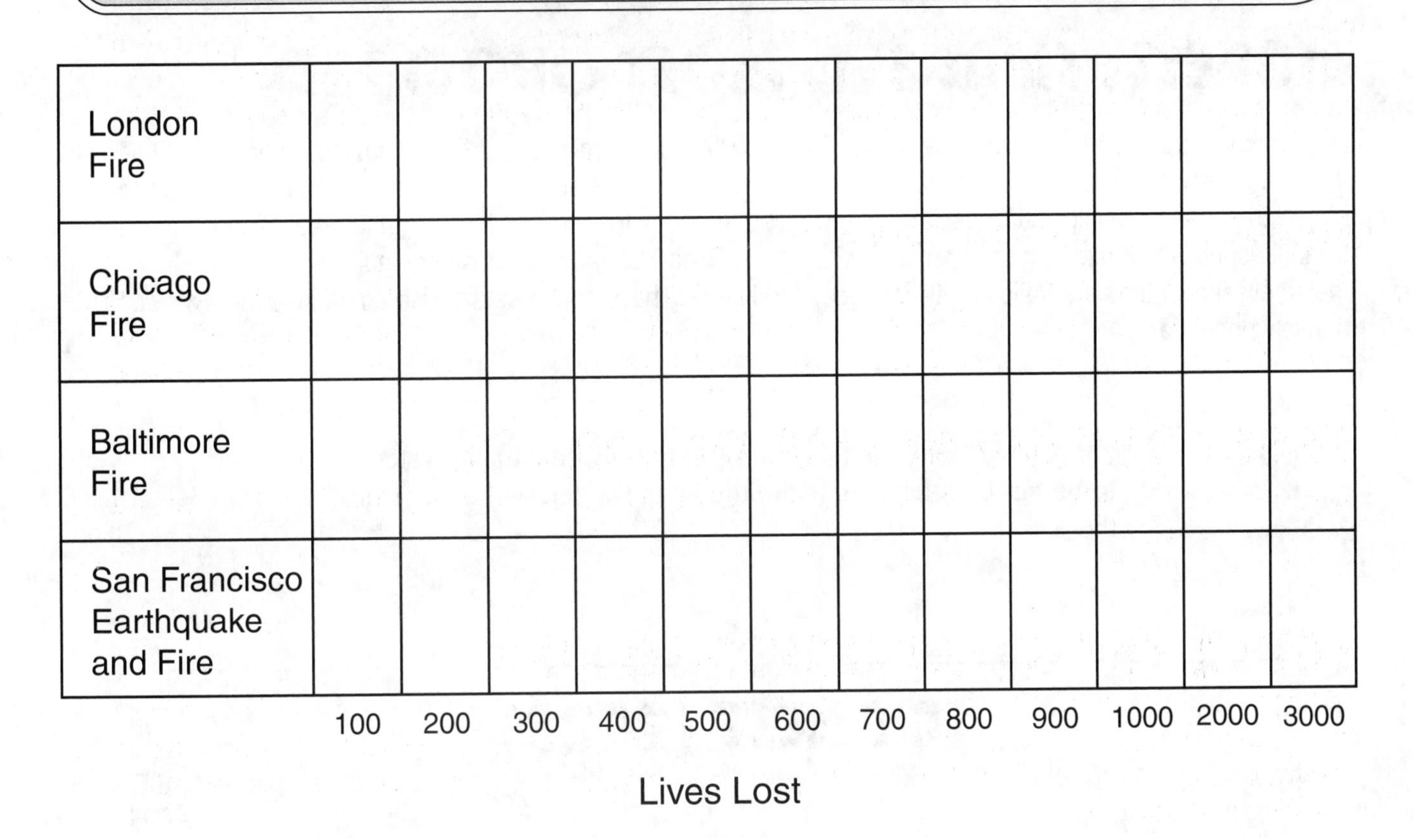

Use the information from Disasters, page 18, to complete the horizontal bar graph. Round your answers to the nearest hundreds place. If there is a death toll that is extremely small, indicate the number by making a small sliver of a line on the graph above.

1. Which fire caused the most deaths? ____________________

2. Which fires had only four deaths? ____________________

3. How many more deaths in the Chicago than in the London fire? ____________________

4. How many more deaths in the San Francisco than in Chicago? ____________________

5. How many lives were lost in all four fires? ____________________

Challenge: Create a vertical bar graph using the death toll from aircraft disasters. Create a graph using increments of 50 for the death toll on the left side of your graph. Create five questions and answers to go with your bar graph.

Smallest Population per Square Mile

BIGGEST, SMALLEST, MOST CROWDED

It's a big world out there! Our planet has about 196,940,000 square miles of surface area. However, over 70 percent of that area is water. The total land area, 57,506,000 square miles, is about 16 times the area of the United States. So, which nation is the biggest? Russia is on top with over 6.5 million square miles of land area. China comes in a distant second with 3.6 million square miles, while the U.S. and Canada are close behind.
The smallest countries are Vatican City, Monaco, and Nauru. These countries are also among the most crowded (densely populated).

*Land areas for countries below do not include inland water. Rankings by total area, including inland water, will differ from these.

SMALLEST (2005)

Germany
Switzerland
Austria
Liechtenstein

	Country	sq mi
1.	Vatican City	0.17
2.	Monaco	0.75
3.	Nauru	8
4.	Tuvalu	10
5.	San Marino	24
6.	Liechtenstein	62
7.	Marshall Islands	70
8.	Saint Kitts and Nevis	101
9.	Maldives	116
10.	Malta	122

Smallest Population per Square Mile *(cont.)*

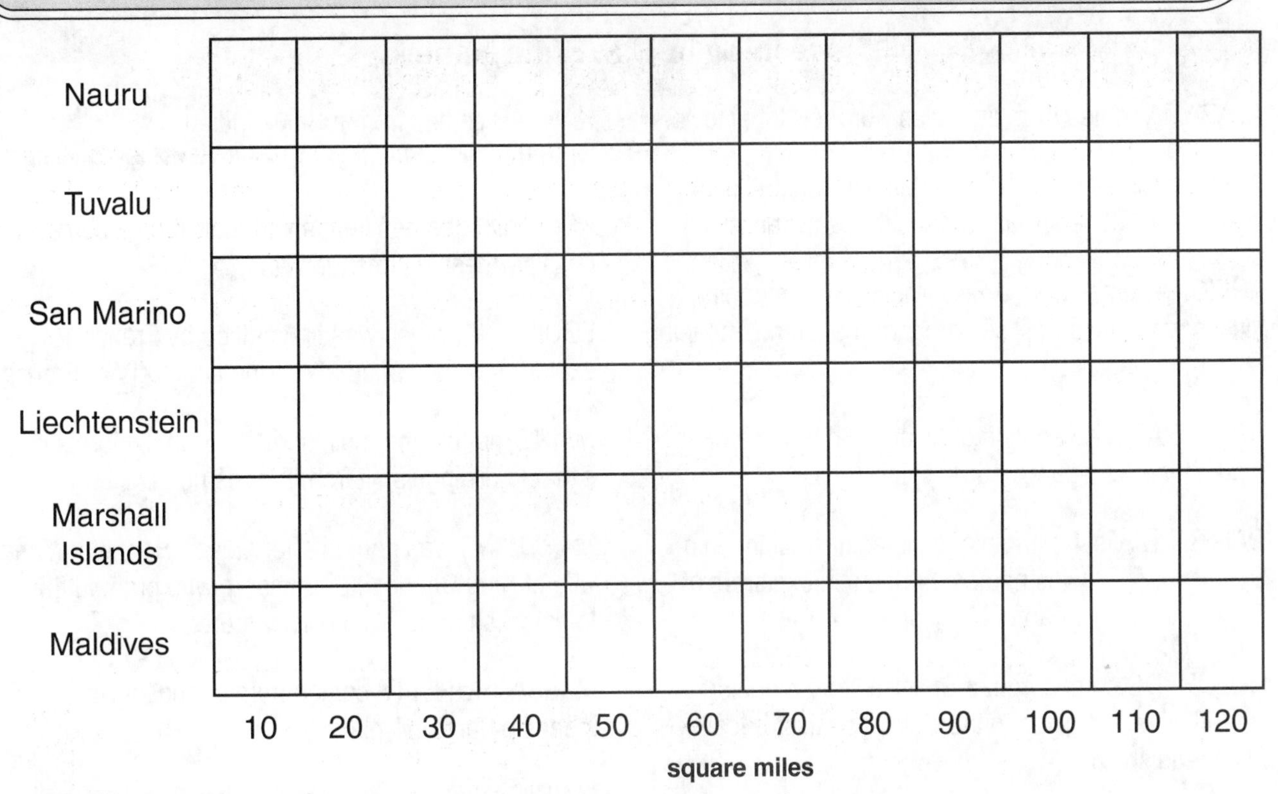

Use information from Smallest Population per Square Mile, page 20, to help you create a horizontal bar graph. Use only the countries listed on the graph. Round numbers of people per square mile to the nearest 10 before graphing.

Use the chart created above to answer the questions below.

1. Which countries have the smallest population per square mile? ______________________

2. Which country has the most people per square mile?______________________

3. Which three countries have less than 25 people per square mile? ______________________

4. How many more people per square mile does Liechtenstein have than Tuvalu? ____________

5. Which country has over 100 people per square mile? ______________________

Challenge: Choose five or more countries from page 20 in another category (densely populated or sparsely populated) and create your own horizontal bar graph. Write five questions and answers for your graph. Discuss your graph with friends.

Number of Letters in the States

Meaning of U.S. State Names

ALABAMA comes from an Indian word for "tribal town."

ALASKA comes from alakshak, the Aleutian (Eskimo) word meaning "peninsula" or "land that is not an island."

ARIZONA comes from a Pima Indian word meaning "little spring place," or the Aztec word arizuma, meaning "silver-bearing."

ARKANSAS is a variation of Quapaw, the name of an Indian tribe. Quapaw means "south wind."

CALIFORNIA is the name of an imaginary island in a Spanish story. It was named by Spanish explorers of Baja California, a part of Mexico.

COLORADO comes from a Spanish word meaning "red." It was first given to the Colorado River because of its reddish color.

CONNECTICUT comes from an Algonquin Indian word meaning "long river place."

DELAWARE is named after Lord De La Warr, the English governor of Virginia in colonial times.

FLORIDA, which means "flowery" in Spanish, was named by the explorer Ponce de León, who landed there during Easter.

GEORGIA was named after King George II of England, who granted the right to create a colony there in 1732.

HAWAII probably comes from Hawaiki, or Owhyhee, the native Polynesian word for "homeland."

IDAHO's name is of uncertain origin, but it may come from a Kiowa Apache name for the Comanche Indians.
ILLINOIS is the French version of Illini, an Algonquin Indian word meaning "men" or "warriors."

INDIANA means "land of the Indians."

IOWA comes from the name of an American Indian tribe that lived on the land that is now the state.

KANSAS comes from a Sioux Indian word that possibly meant "people of the south wind."

KENTUCKY comes from an Iroquois Indian word, possibly meaning "meadowland."

LOUISIANA, which was first settled by French explorers, was named after King Louis XIV of France.

MAINE means "the mainland." English explorers called it that to distinguish it from islands nearby.

MARYLAND was named after Queen Henrietta Maria, wife of King Charles I of England, who granted the right to establish an English colony there.

MASSACHUSETTS comes from an Indian word meaning "large hill place."

MICHIGAN comes from the Chippewa Indian words mici gama, meaning "great water" (referring to Lake Michigan).

MINNESOTA got its name from a Dakota Sioux Indian word meaning "cloudy water" or "sky-tinted water."

MISSISSIPPI is probably from Chippewa Indian words meaning "great river" or "gathering of all the waters," or from an Algonquin word, messipi.

MISSOURI comes from an Algonquin Indian term meaning "river of the big canoes."

Number of Letters in the States *(cont.)*

Informational Chart: (record the number of letters in each state's name)

4	5	6	7	8	9	10	11	12	13
___	___	___	___	___	___	___	___	___	___
25	25	25	25	25	25	25	25	25	25

Example: Alabama = 7 letters

Use all 25 of the states on Number of letters in the States, page 22. Count the number of letters in each name. Make a circle graph to show the percentage of letters in these states.

Use the information above to create a circle graph below. Use the fraction above to help you find the percentage to help you create the circle (pie) graph. Multiply the numerator by 100 and divide that answer by the denominator (25).

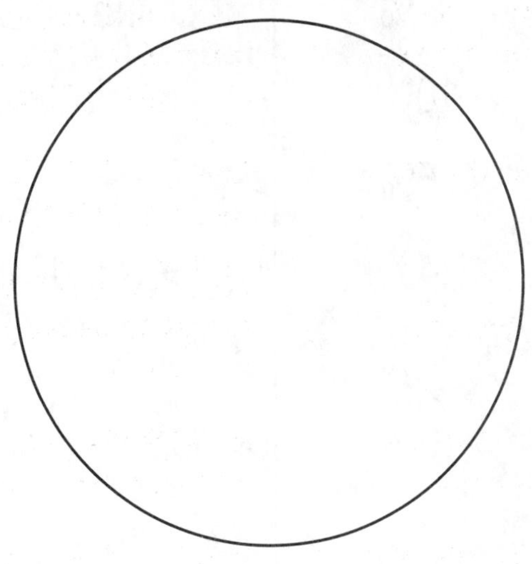

Answer the following questions using the circle graph above.

1. What percentage begin with nine letters? ______________________________
2. What percentage begin with twelve letters? ______________________________
3. How many more percent begin with eight letters that six letters? ______________________________
4. What percent begin with five and six letters combined? ______________________________
5. If you combine the two largest groups of letters, what is the percentage? ______________________________

Challenge: Create a bar graph to show the same information using the other 25 states. Write five questions and answers to the circle graph.

Top Energy Users

WHO PRODUCES AND USES THE MOST ENERGY?

The United States produces about 18% of the world's energy—more than any other country—but it also uses 24% of the world's supply. The table below on the left lists the world's top ten energy producers and the percent of the world's production that each nation was responsible for in 2002. The table on the right lists the world's top energy users and the percent of the world's energy that each nation consumed.

TOP ENERGY PRODUCERS		TOP ENERGY USERS	
United States	18%	United States	24%
Russia	11%	China	10%
China	10%	Russia	7%
Saudi Arabia	5%	Japan	5%
Canada	4%	Germany	3%
Australia	3%	India	3%
India	3%	Canada	3%
Iran	3%	France	2%
Norway	3%	United Kingdom	2%
United Kingdom	3%	Brazil	2%

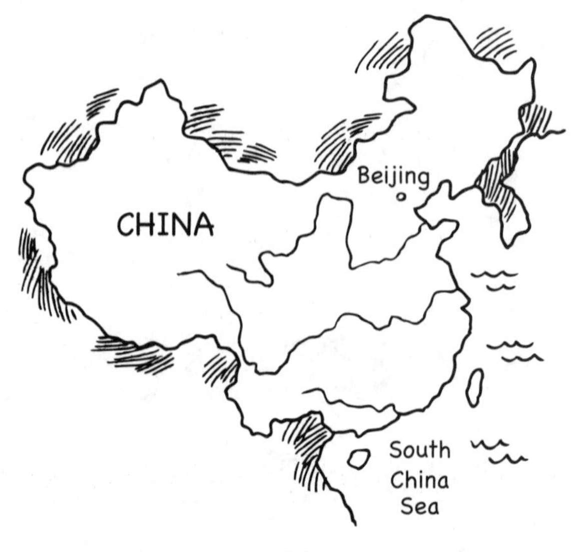

Top Energy Users *(cont.)*

Read the information from Top Energy Users, page 24. Use the information to create a circle graph. The percentages have already been given, so the job will be to find out the missing percentage which will be the amount of energy the rest of the world uses. The number on the bottom (denominator) of your fraction will be 100.

Use the circle (pie) graph above to answer the following questions:

1. Which country uses the most energy?______________________________

2. Which countries use the least amount of energy?______________________________

3. How much energy do China and Russia use together?______________________________

4. How much more energy does the United States use than Germany? ______________________________

5. Look at a map of the world, list two or more reasons why the U.S. might use so much energy. ______________________________

____

Challenge: Make a second circle graph which shows the top energy producers of the world. Write five questions and answers for your new graph.

Where Garbage Goes

WHERE GARBAGE GOES

Most of the things around you will be thrown away someday. Skates, clothes, the toaster, furniture—they can break or wear out, or you may get tired of them. Where will they go when they are thrown out? What kinds of waste will they create?

LOOK AT WHAT IS NOW IN U.S. LANDFILLS

Material		Percent
Metal	⟶	8%
Plastic	⟶	24%
Food and Yard Waste	⟶	11%
Rubber and Leather	⟶	6%
Other Trash	⟶	21%
Paper	⟶	30%

What Happens to Things We Throw Away?

Landfills

Most of our trash goes to places called landfills. A landfill (or dump) is a low area of land that is filled with garbage. Most modern landfills are lined with a layer of plastic or clay to try to keep dangerous liquids from seeping into the soil and ground water supply.

The Problem with Landfills

More than half of the states in this country are running out of places to dump their garbage. Because of the unhealthful materials many contain, landfills do not make good neighbors, and people don't want to live near them. But where can cities dispose of their waste? How can hazardous waste — material that can poison air, land, and water — be disposed of in a safe way?

Incinerators

One way to get rid of trash is to burn it. Trash is burned in a furnace-like device called an incinerator. Because incinerators can get rid of almost all of the bulk of the trash, some communities would rather use incinerators than landfills.

The Problem with Incinerators

Leftover ash and smoke from burning trash may contain harmful chemicals, called pollutants, and make it hard for some people to breathe. They can harm plants, animals, and people.

That's a lot of paper! The U.S. recycled an all-time high of 50 million tons of paper, or about 300 pounds per person, in 2004. That's about half of the paper produced in the U.S. in a year. By the year 2012, the paper industry hopes to recycle 55% of the paper it produces. Germany and Finland currently reuse about 75% of the paper they produce. Back in 1990, the U.S. was recycling only about a third of its paper.

Where Garbage Goes *(cont.)*

Read the information Where Garbage Goes, page 26. Use it to create a circle graph showing the percentage of different types of things which can be found in the landfills. The number to use on the bottom (denominator) of your fraction will be 100.

Answer the following questions using the circle (pie) graph above.

1. What is the percentage of metal and plastic found in the landfill? ____________________

2. If you were to take all the paper out of the landfill, what percentage of room would be available for something else? ____________________

3. What is the second largest percentage of items discarded? ____________________

4. How much miscellaneous trash is found in the landfill? ____________________

5. Which area does your family contribute to the most? ____________________

Challenge: Create a circle graph using data from all of the students in your class. Record information about the number of siblings they have. Make a circle graph to show the percentage of siblings your class has. Remember, the number on the bottom (denominator) of the fraction will be the total number of children in your classroom.

The Growing U.S. Population

THE GROWING U.S. POPULATION

1790: 3,929,214
1850: 23,191,876
1900: 76,212,168
1930: 123,202,624
1950: 151,325,798
1970: 203,302,031
1990: 248,709,873
2000: 281,421,906

The Growing U.S. Population *(cont.)*

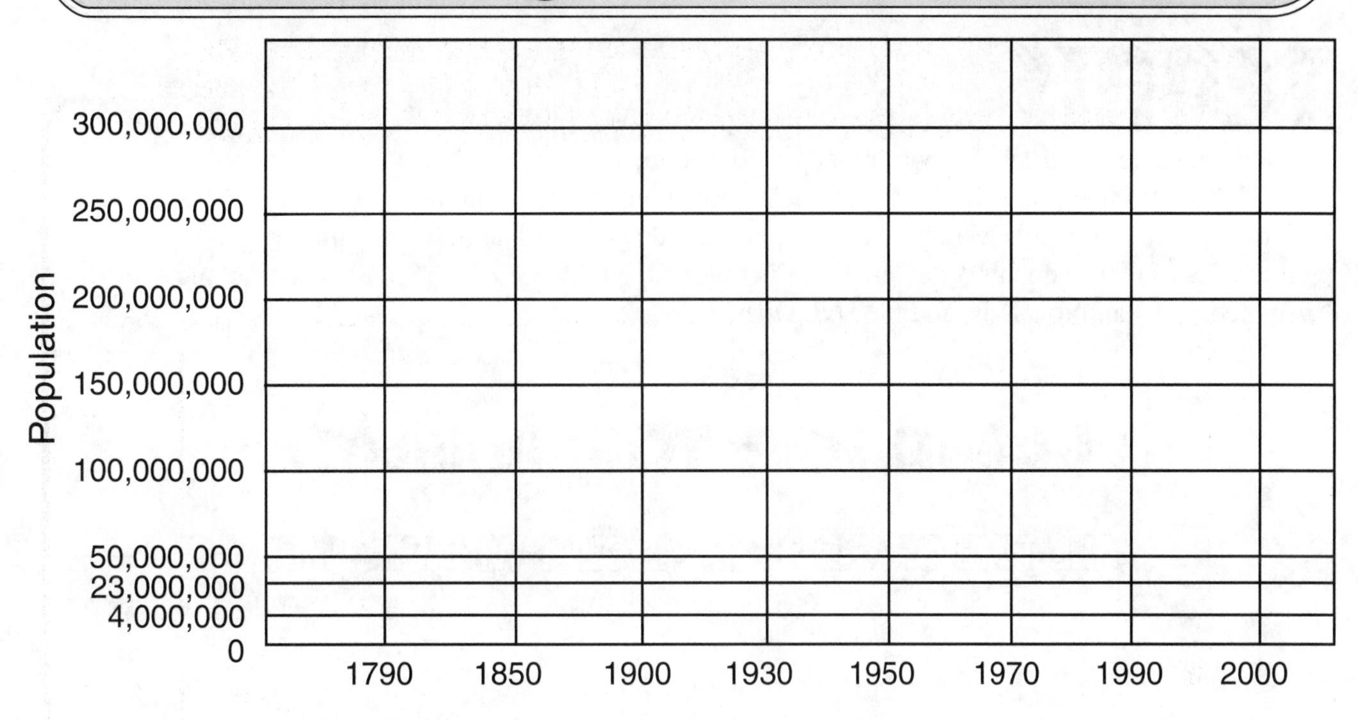

Use the information from The Growing U.S. Population, page 28. Use the information to create your line graph about the rising population. Round all numbers under 50 million to the nearest one million. Round all numbers over 50 million to the nearest 50 million.

Use the line graph to answer the questions below:

1. What was the population of the U.S. in the year 1790? ______________________________

2. How much did the population grow between 1790 and 1850?______________________________

3. What is the difference in population in 1970 from the population in the year 2000? __________

4. How much has the population grown between 1930 and 1950?______________________________

5. What do you think is one of the main reasons the population of the United States continues to grow?

____

Challenge: Collect information about population growth over the past 100 years in the city or town you live in. Create a line graph to record your findings. Write five questions and answers to accompany your graph.

Grand Slams

TENNIS

Modern tennis began in 1873. It was based on court tennis.
In 1877 the first championships were held in Wimbledon, near London. In 1881 the first official U.S. men's championships were held at Newport, Rhode Island. The four most important ("Grand Slam") tournaments today are the Australian Open the French Open, the All-England (Wimbledon) Championships, and the U.S. Open.

GRAND SLAM TOURNAMENTS

ALL-TIME GRAND SLAM SINGLES WINNERS–MEN

	Australian	French	Wimbledon	U.S.	Total
Pete Sampras (b. 1971)	2	0	7	5	14
Roy Emerson (b. 1936)	6	2	2	2	12
Bjorn Borg (b. 1956)	0	6	5	0	11
Rod Laver (b. 1938)	3	2	4	2	11
Bill Tilden (1893–1953)	*	0	3	7	10

Grand Slams *(cont.)*

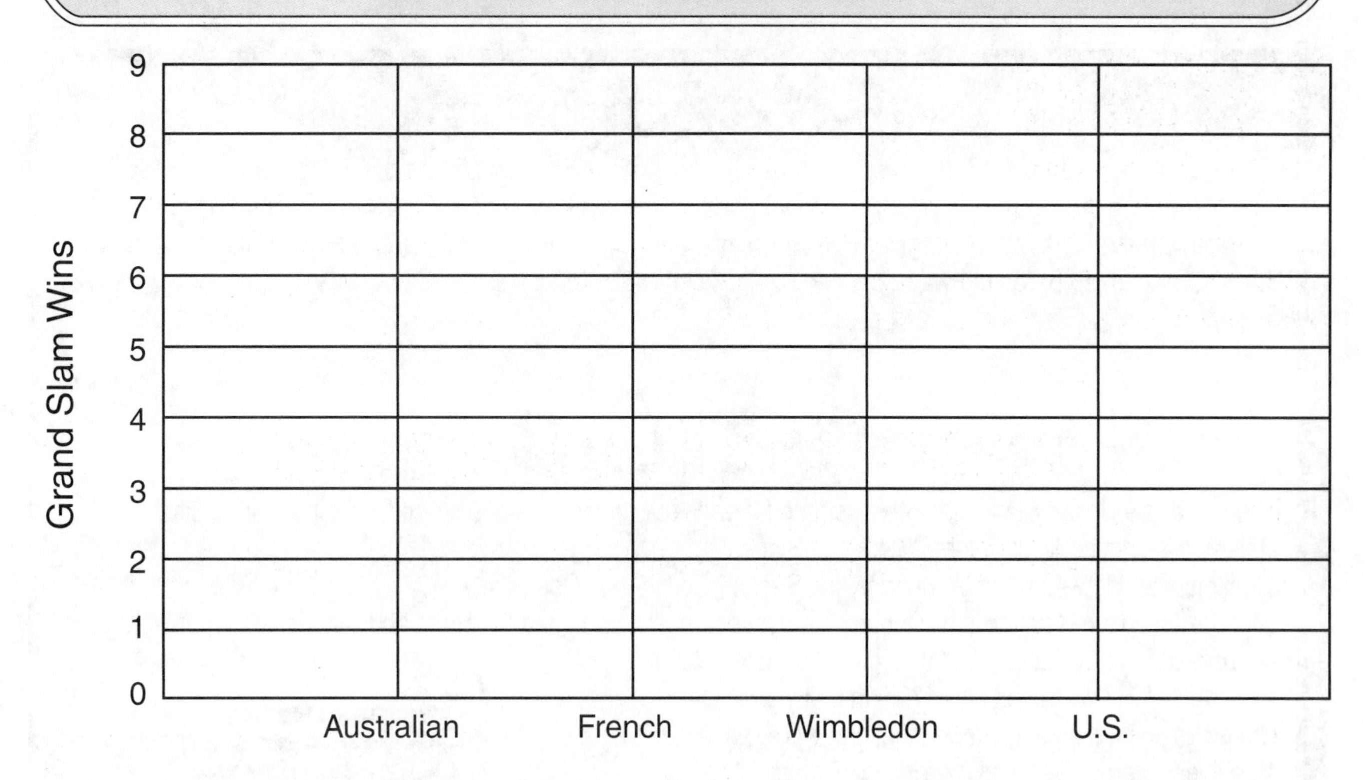

Use the information from Grand Slam Tournaments, page 30, to create a line graph about Pete Sampras' wins.

Use the line graph above to answer questions about the four tournaments.

1. How may grand slams did he win in the Australian tournament? ____________________

2. How many more did he win in the Wimbledon tournament than in the French? ____________

3. How many Grand Slam wins did he have with the U.S. and Australian tournaments combined?

   ____

4. How many more Grand Slam wins did he get in Wimbledon than in the Australian tournament?

   ____

5. If he only received two Grand Slams wins in the U.S. tournament instead of five, what would his grand total have been?

   ____

Challenge: Create other line graphs to show the tournaments for other tennis players. Compare different tennis players; write five questions and answers for each line graph you create. (Or, make the information into a multiple line graph using several different colors to represent several different players. Make sure to include a key to identify which color represents which player.)

Symbols of the United States

The Great Seal

The Great Seal of the United States shows an American bald eagle with a ribbon in its mouth bearing the Latin words e *pluribus unum* (out of many, one). In its talons are the arrows of war and an olive branch of peace. On the back of the Great Seal is an unfinished pyramid with an eye (the eye of Providence) above it. The seal was approved by Congress on June 20, 1782.

THE FLAG

The flag of the United States has 50 stars (one for each state) and 13 stripes (one for each of the original 13 states). It is unofficially called the "Stars and Stripes."

The first U.S. flag was commissioned by the Second Continental Congress in 1777 but did not exist until 1783, after the American Revolution. Historians are not certain who designed the Stars and Stripes. Many different flags are believed to have been used during the American Revolution.

The flag of 1777 was used until 1795. In that year Congress passed an act ordering that a new flag have 15 stripes, alternate red and white, and 15 stars on a blue field. In 1818, Congress directed that the flag have 13 stripes and that a new star be added for each new state of the Union. The last star was added in 1960 for the state of Hawaii.

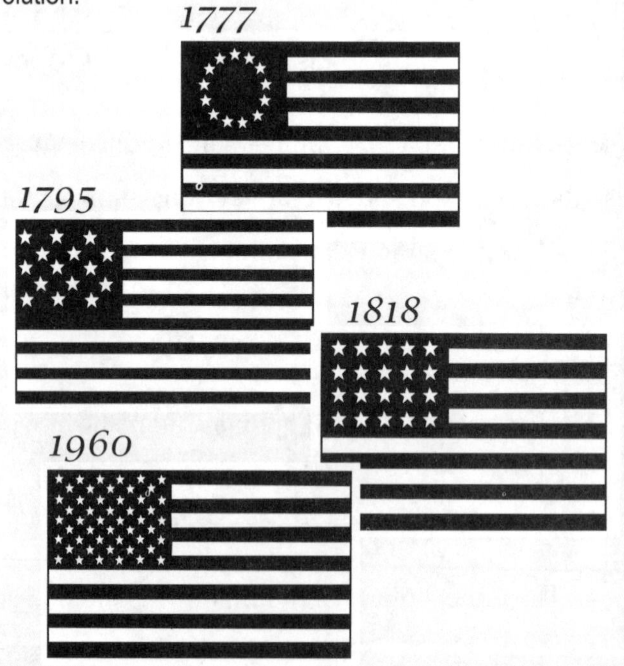

There are many customs for flying the flag and treating it with respect. For example, it should not touch the floor and no other flag should be flown above it, except for the UN flag at UN head-quarters. When the flag is raised or lowered, or passes in a parade, or during the Pledge of Allegiance, people should face it and stand at attention. Those in military uniform should salute. Others should put their right hand over their heart. The flag is flown at half staff as a sign of mourning.

Pledge of Allegiance to the Flag

"I pledge allegiance to the flag of the United States of America and to the republic for which it stands, one nation under God, indivisible, with liberty and justice for all."

THE NATIONAL ANTHEM

"The Star-Spangled Banner" was a poem written in 1814 by Francis Scott Key as he watched British ships bombard Fort McHenry, Maryland, during the War of 1812. It became the National Anthem by an act of Congress in 1931. The music to "The Star-Spangled Banner" was originally a tune called "Anacreon in Heaven."

Symbols of the United States *(cont.)*

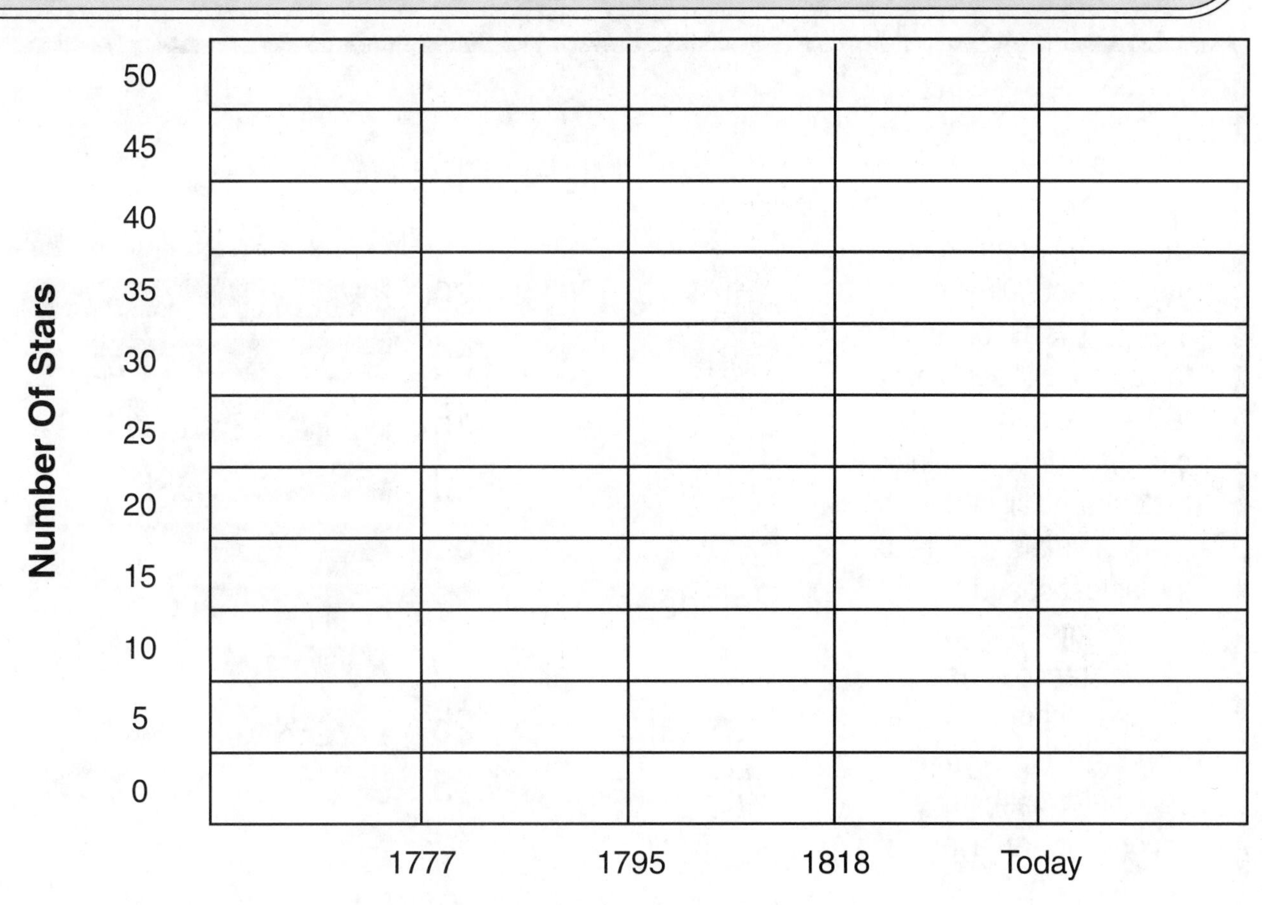

Use the information from Symbols of the United States, page 32, to create a line graph and to discover how the number of stars on the flag has changed throughout the course of history. Round the numbers before graphing.

Use the line graph above to answer the questions below.

1. How many stars are on the American Flag today? ________________________

2. How many stars were on the original flag? ________________________

3. Compare the flag in 1818 to today's flag. What is the difference? ________________________

4. How many stars were added between 1795 and 1818? ________________________

5. How many stars were added between 1777 and 1795? ________________________

Challenge: Check the temperature in a city in the United States over the past four days. Create a line graph to show the highs or the lows for the past week. Discuss your findings. Then write five questions and answers to accompany the line graph.

How Fast Do Animals Run?

Some animals can run as fast as a car can drive. But a snail needs more than 30 hours just to go 1 mile. If you look at this table, you will see how fast some land animals can go. Humans at their fastest are still slower than many animals. The record for fastest speed for a human for a recognized race distance is held by Michael Johnson, who won the 1996 Olympic 200-meter dash in 19.32 seconds for an average speed of 23.16 mph.

	MILES PER HOUR
Cheetah	70
Antelope	60
Lion	50
Elk	45
Zebra	40
Rabbit	35
Reindeer	32
Cat	30
Elephant	25
Wild turkey	15
Squirrel	12
Snail	0.03

How Fast Do Animals Run? *(cont.)*

70

60

50

40

30

20

10

antelope cheetah cat rabbit wild turkey elk lion

Use the information on How Fast Do Animals Run?, page 34, to create a line plot that shows the miles per hour the particular animals listed on the line plot above can run. Round the speed of each animal to the nearest 10 and use one X to represent 10 miles per hour.

Use the line plot above to answer the questions below:

1. What is the speed of the fastest animal listed on the line plot? ____________________
2. What is the difference in speed between a rabbit and an elk? ____________________
3. Which animals run as quickly as or more quickly than a car driving (about 60 miles per hour)?

   ____
4. Which animal runs slower than 20 miles per hour? ____________________
5. How much faster does a cheetah run than a lion? ____________________

Challenge: Choose five other animals that are not on the line plot or research other animals not listed on the chart on page 34. Create a line plot to show the speed each animal is capable of running. Make sure to identify what X is equal to. Include five questions and answers to accompany your line plot.

Kinds of Musical Instruments

There are many kinds of musical instruments. Instruments in an orchestra are divided into four groups, or sections: string, woodwind, brass, and percussion.

PERCUSSION INSTRUMENTS

Percussion instruments make sounds when they are struck. The most common percussion instrument is the drum. Others include cymbals, triangles, gongs, bells, and xylophone. Keyboard instruments, like the piano, are sometimes thought of as percussion instruments.

BRASSES

Brass instruments are hollow inside. They make sounds when air is blown into a mouthpiece shaped like a cup or a funnel. The trumpet, French horn, trombone, and tuba are brasses.

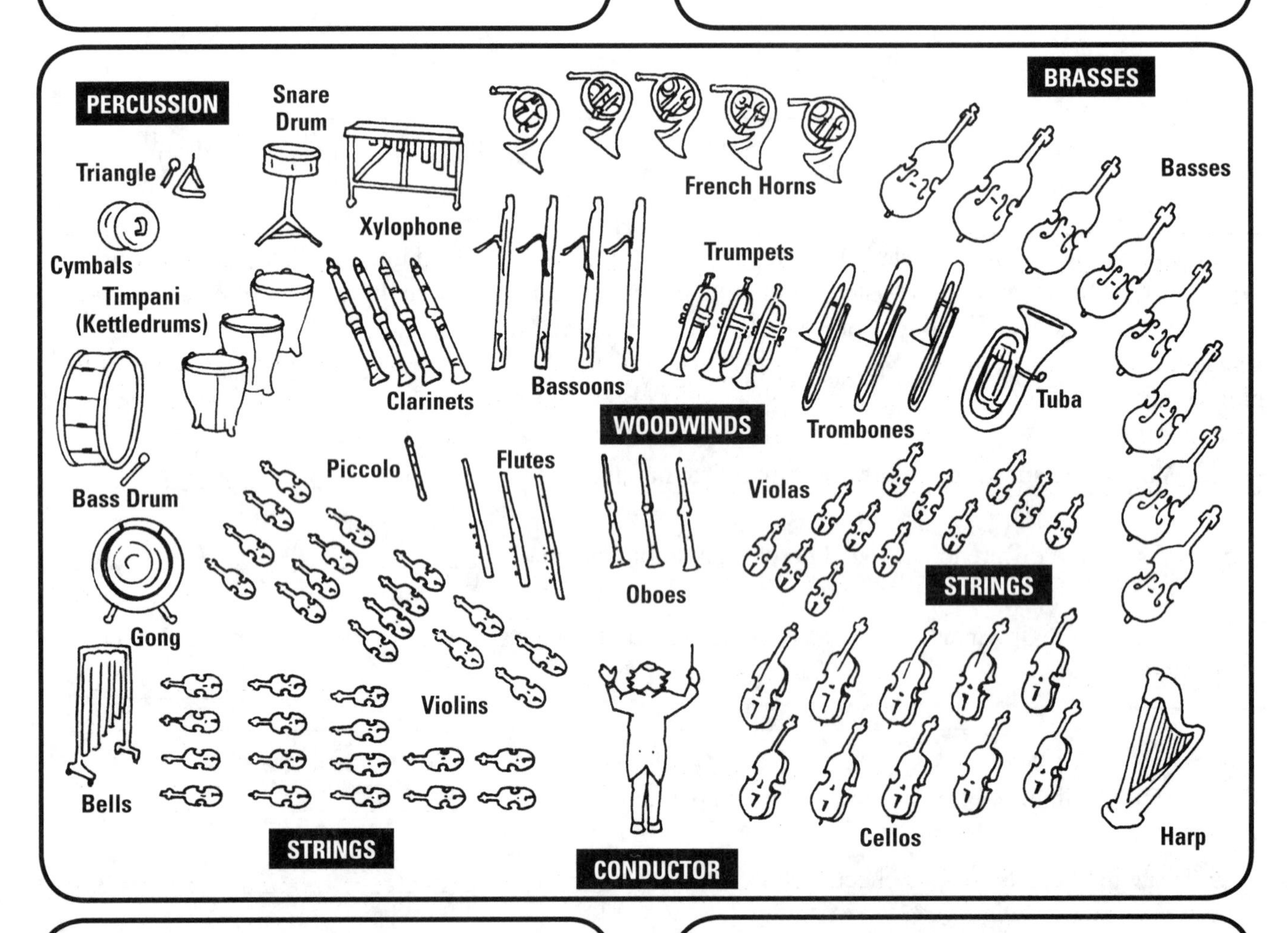

WOODWINDS

Woodwind instruments are long and round and hollow inside. They make sounds when air is blown into them through a mouth hole or a reed. The clarinet, flute, oboe, bassoon, and piccolo are woodwinds.

STRINGS

Stringed instruments make sounds when the strings are either stroked with a bow or plucked with the fingers. The violin, viola, cello, bass, and harp are used in an orchestra. The guitar, banjo, and mandolin are other stringed instruments.

Kinds of Musical Instruments *(cont.)*

French horn cello violins trombone bass oboe flute kettledrum

Use Kinds of Musical Instruments, page 36, to discover the world of musical instruments. Create a line plot to show how many people are playing each instrument in the orchestra. Use one X to symbolize each instrument. Use only the instruments listed above to complete the line plot.

Use the line plot above to help you answer the questions below:

1. Which instrument has the largest number on the line plot above? ____________________

2. Which instruments have the least number on the line plot above? ____________________

3. How many more cellos than oboes are there? ____________________

4. How many fewer flutes are in the orchestra than basses? ____________________

5. How many instruments would there be with the French horns and violas combined? __________

Challenge: Count the number of pencils in the desks of 10 people in your classroom. Make a line plot to show the number of pencils each one has. Use one X for each pencil. Write five questions and answers to use in conjunction with the line plot you created.

Record Temperatures

COLDEST TEMPERATURE: Alaska

HOTTEST TEMPERATURE: California

(THROUGH 2000)

State	LOWEST ° F	LOWEST Latest date	HIGHEST ° F	HIGHEST Latest date
Alabama	–27	Jan. 30, 1966	112	Sept. 5, 1925
Alaska	–80	Jan. 23, 1971	100	June 27, 1915
Arizona	–40	Jan. 7, 1971	128	June 29, 1994
Arkansas	–29	Feb. 13, 1905	120	Aug. 10, 1936
California	–45	Jan. 20, 1937	134	July 10, 1913
Colorado	–61	Feb. 1, 1985	118	July 11, 1888
Connecticut	–32	Jan. 22, 1961	106	July 15, 1995
Delaware	–17	Jan. 17, 1893	110	July 21, 1930
Florida	–2	Feb. 13, 1899	109	June 29, 1931
Georgia	–17	Jan. 27, 1940	112	Aug. 20, 1983
Hawaii	12	May 17, 1979	100	Apr. 27, 1931
Idaho	–60	Jan. 18, 1943	118	July 28, 1934
Illinois	–36	Jan. 5, 1999	117	July 14, 1954
Indiana	–36	Jan. 19, 1994	116	July 14, 1936
Iowa	–47	Feb. 3, 1996	118	July 20, 1934
Kansas	–40	Feb. 13, 1905	121	July 24, 1936
Kentucky	–37	Jan. 19, 1994	114	July 28, 1930
Louisiana	–16	Feb. 13, 1899	114	Aug. 10, 1936
Maine	–48	Jan. 19, 1925	105	July 10, 1911
Maryland	–40	Jan. 13, 1912	109	July 10, 1936
Massachusetts	–35	Jan. 12, 1981	107	Aug. 2, 1975
Michigan	–51	Feb. 9, 1934	112	July 13, 1936
Minnesota	–60	Feb. 2, 1996	114	July 6, 1936
Mississippi	–19	Jan. 30, 1966	115	July 29, 1930
Missouri	–40	Feb. 13, 1905	118	July 14, 1954
Montana	–70	Jan. 20, 1954	117	July 5, 1937
Nebraska	–47	Dec. 22, 1989	118	July 24, 1936
Nevada	–50	Jan. 8, 1937	125	June 29, 1994
New Hampshire	–47	Jan. 29, 1934	106	July 4, 1911
New Jersey	–34	Jan. 5, 1904	110	July 10, 1936
New Mexico	–50	Feb. 1, 1951	122	June 27, 1994
New York	–52	Feb. 18, 1979	108	July 22, 1926
North Carolina	–34	Jan. 21, 1985	110	Aug. 21, 1983
North Dakota	–60	Feb. 15, 1936	121	July 6, 1936
Ohio	–39	Feb. 10, 1899	113	July 21, 1934
Oklahoma	–27	Jan. 18, 1930	120	June 27, 1994
Oregon	–54	Feb. 10, 1933	119	Aug. 10, 1898
Pennsylvania	–42	Jan. 5, 1904	111	July 10, 1936
Rhode Island	–25	Feb. 5, 1996	104	Aug. 2, 1975
South Carolina	–19	Jan. 21, 1985	111	June 28, 1954
South Dakota	–58	Feb. 17, 1936	120	July 5, 1936
Tennessee	–32	Dec. 30, 1917	113	Aug. 9, 1930
Texas	–23	Feb. 8, 1933	120	June 28, 1994
Utah	–69	Feb. 1, 1985	117	Jul. 5, 1985
Vermont	–50	Dec. 30, 1933	105	July 4, 1911
Virginia	–30	Jan. 22, 1985	110	July 15, 1954
Washington	–48	Dec. 30, 1968	118	Aug. 5, 1961
West Virginia	–37	Dec. 30, 1917	112	July 10, 1936
Wisconsin	–55	Feb. 4, 1996	114	July 13, 1936
Wyoming	–66	Feb. 9, 1933	115	Aug. 8, 1983

Record Temperatures *(cont.)*

135°							
130°							
125°							
120°							
115°							
110°							
105°							
100°							
	California	Florida	South Carolina	South Dakota	Wisconsin	Wyoming	Minnesota

Use Record Temperatures, page 38, to gather information for the line plot. Use each X to symbolize five degrees. Use only the states represented on the line plot above. Round each number to the nearest 5.

Use the line plot above to answer the questions below.

1. Which state was the hottest place to visit, due to the temperature? ____________________

2. If you wanted to visit the two coolest states, which ones would they be? ____________________

3. How many degrees warmer is it in Californian than in Wyoming? ____________________

4. What is the difference between South Carolina and South Dakota's heat? ____________________

5. How much cooler is Florida than Wisconsin? ____________________

Challenge: Create your own line plot to show weather in seven other states. Make sure to identify what the value of X will be on your line plot. Write five questions and answers to accompany your line plot.

Dinosaur Celebrities

Apatosaurus

Deceptive lizard
Plant-eating
Length: 70+ feet
Period: Jurassic
Found in: Western U.S.

Camptosaurus

Bent lizard • Plant-eating
Length: 20 feet • Period: Jurassic-Cretaceous
Found in: North America and Western Europe

Hadrosaurus

Big lizard
Plant-eating
Length: 30 feet
Period: Cretaceous
Found in: Asia, Europe, North and South America

Triceratops

Three-horned face • Plant-eating Length: 30 feet • Period: Cretaceous Found in: North America

Stegosaurus

Plated lizard • Plant-eating • Length: 30 feet • Period: Jurassic • Found in: North America

Velociraptor

Speedy thief
Meat-eating
Length: 6 feet
Period: Cretaceous
Found in: Asia

Tyrannosaurus Rex ("T-Rex")

King of the tyrant lizards
Meat-eating • Length: 40 feet
Period: Cretaceous
Found in: Western U.S., Canada, Asia

Dinosaur Celebrities *(cont.)*

Hadrosaurus Apatosaurus Tyrannossaurus Stegosaurus Camptosaurus

Use then information from Dinosaur Celebrities, page 40, to gather information about the lengths of the dinosaurs of the past. Use one X on the line plot to symbolize 10 feet in length.

Use the line plot above to answer the questions below:

1. Which dinosaur is the shortest in length? ______________________

2. Which dinosaur is the longest? ______________________

3. How much shorter is Camptosaurus than Stegosaurus? ______________________

4. How much longer is Tyrannosaurus than Hadrosaurus? ______________________

5. If Apatosaurus and Camptosaurus were side by side, who would be shorter? ______________________

Challenge: Measure the length of 10 objects in your classroom using inches. Record your findings on a line plot. Identify the value of X on your line plot. Write five questions and answers to accompany your line plot.

Largest Cities in the World

Largest Ten Cities in the World

Here are the ten cities that had the most people, according to UN estimates for 2003. Numbers include people from the built-up area around each city (metropolitan area), not just the city.

	City, Country	Population
1.	Tokyo, Japan	34,997,300
2.	Mexico City, Mexico	18,660,200
3.	New York area, U.S.	18,252,300
4.	São Paulo, Brazil	17,857,000
5.	Mumbai (Bombay), India	17,431,300
6.	Delhi, India	14,146,000
7.	Calcutta, India	13,805,700
8.	Buenos Aires, Argentina	13,047,000
9.	Shanghai, China	12,759,000
10.	Jakarta, Indonesia	12,296,000

Largest Cities in the World *(cont.)*

Buenos Aires | Sao Paulo | New York | Calcutta | Tokyo | Shanghai

Use the information from Largest 10 Cities in the World, page 42, to gather information about the largest cities of the world. Use one X on the line plot below to symbolize one million people. Round each of the numbers to the nearest million before plotting it.

Use the line plot above to answer the questions below.

1. Which city has the largest population? ______________________
2. Which cities have the smallest population? ______________________
3. How many more people live in San Paulo than Buenos Aires? ______________________
4. How many fewer people live in Calcutta than New York? ______________________
5. If you added Tokyo, Shanghai, and Calcutta, what would be the population? ______________________

Challenge: Take a survey of 10 friends on the playground. Ask how many pets they have at home. Create a line plot to show the results for each friend. (Make sure to record their names and the number of pets they have so your results will be accurate on your line plot!) X on this line plot should represent one pet. Write five questions and answers to accompany the line plot.

TENNIS

Modern tennis began in 1873. It was based on court tennis.
In 1877 the first championships were held in Wimbledon, near London. In 1881 the first official U.S. men's championships were held at Newport, Rhode Island. Six years later, the first U.S. women's championships took place, in Philadelphia. The four most important ("grand slam") tournaments today are the Australian Open, the French Open, the All-England (Wimbledon) Championships, and the U.S. Open.

GRAND SLAM TOURNAMENTS

ALL-TIME GRAND SLAM SINGLES WINNERS

MEN	Australian	French	Wimbledon	U.S.	Total
Pete Sampras (b. 1971)	2	0	7	5	14
Roy Emerson (b. 1936)	6	2	2	2	12
Bjorn Borg (b. 1956)	0	6	5	0	11
Rod Laver (b. 1938)	3	2	4	2	11
Bill Tilden (1893–1953)	*	0	3	7	10
WOMEN					
Margaret Smith-Court (b. 1942)	11	5	3	5	24
Steffi Graf (b. 1969)	4	6	7	5	22
Helen Wills Moody (1905–1998)	*	4	8	7	19
Chris Evert (b. 1954)	2	7	3	6	18
Martina Navratilova (b. 1956)	3	2	9	4	18

*Never played in tournament.

Tennis Tournaments *(cont.)*

Rod Laver	Bill Tilden	Steffi Graf	Chris Evert	Martina Navatilova	Bjorn Borg

Use the information from Tennis Tournaments, page 44. Use one X on the line plot to symbolize one win. Use the total number of tournaments won to gather your information.

Use the line plot above to answer the questions below.

1. Which player has won the most tournaments? ______________________________
2. Which player has won the least? ______________________________
3. How many more tournaments has the top player won than the last player? ______________
4. How many tournaments have all the players won in all? ______________________
5. Who holds the most tournaments won combined? The men or the women? ______________

Challenge: Ask 10 friends the number of vehicles their families have at home. Create a line plot to show the information gathered. Use one X to represent one vehicle. Make sure to write five questions and answers for your line plots.

Exercise—It's What You Do!

If you watch TV in the afternoons after school or on Saturday mornings, you've probably seen the "VERB. It's What You Do" ads. They're part of a seven-year campaign sponsored by the government to encourage kids age 9-13 to get more exercise.

Why does the U.S. Department of Health and Human Services' Centers for Disease Control and Prevention (CDC) think exercise is important for kids? In 2002, the National Center for Health Statistics reported that an estimated 8.8 million U.S. kids age 6-19 were overweight. Being overweight increases a risk of developing high blood pressure, diabetes, and heart disease.

But daily exercise has other benefits, too: it makes you feel good. Exercise also helps you think better, sleep better, and feel more relaxed. Regular exercise will make you stronger and help you improve at physical activities. Breathing deeply during exercise gets more oxygen into your lungs with each breath. Your heart pumps more oxygen-filled blood all through your body with each beat. Muscles and joints get stronger and more flexible as you use them.

Organized sports are a good way to get a lot of exercise, but not the only way. You can shoot hoops, jog, ride a bike, or skate without being on a team. If you can't think of anything else, try walking in a safe place. Walk with friends or even try to get the adults in your life to join you. They could probably use the exercise, too!

Below are some activities, with a rough idea of how many calories a 100-pound person would burn per minute while doing them.

Activity	Calories Per Minute
Jogging (6 miles per hour)	8
Jumping rope (easy pace)	7
Playing basketball	7
Playing soccer	6
Bicycling (9.4 miles per hour)	5
Skiing (downhill)	5
Raking the lawn	4
Rollerblading (easy pace)	4
Walking (4 miles per hour)	4
Bicycling (5.5 miles per hour)	3
Swimming (25 yards per minute)	3
Walking (3 miles per hour)	3

Exercise—It's What You Do! *(cont.)*

Bicycling (9.4 mph)	
Bicycling (5.5 mph)	
Jogging (6 mph)	
Jumping rope (easy pace)	
Playing basketball	
Playing soccer	
Rollerblading	
Skiing (downhill)	
Swimming (25 ypm)	
Walking (3mph)	

Use the information from Exercise—It's What You Do, page 46, to assist in gathering data for the picture graph. One ☺ will equal one calorie per minute.

Use the picture graph above to answer the following questions:

1. Which activity or activities will help you burn the least amount of calories per minute?

   ____

2. Which biking activity would be the best calorie burner. . . 5.5 miles per hour or 9.4?

   ____

3. Which activity or activities would you choose if you wanted to burn about 4 calories a minute?

   ____

4. Which two activities would burn calories at seven calories per minute? ______________

   ____

5. How many more calories would you burn per minute if you were jumping rope instead of playing soccer?

   ____

Challenge: Choose six of the sports above. Ask 10 students to choose which activity they would prefer to do, if they could only choose one. Create a picture graph to record the information gathered. Make sure to include five questions and answers with the picture graph.

Facts About Nations

Tunisia

✦Capital: Tunis
✦Population: 10,074,951*
✦Area: 63,170 sq. mi. (163,610 sq. km.)
✦Currency: $1 = 1.24 dinars
✦Language: Arabic, French
✦Did You Know: With more than 800 miles of coastline on the Mediterranean Sea, Tunisia is a popular vacation spot for European tourists.

Uganda

✦Capital: Kampala
✦Population: 27,269,482
✦Area: 91,136 sq. mi. (236,040 sq. km.)
✦Currency: $1 = 1,712 shillings
✦Language: English, Luganda, Swahili
✦Did You Know: Bwindi Impenetrable National Park is a home to endangered mountain gorillas.

Turkey

✦Capital: Ankara
✦Population: 69,660,559
✦Area: 301,383 sq. mi. (780,580 sq. km.)
✦Currency: $1 = 1.34 new lira
✦Language: Turkish, Kurdish, Arabic
✦Did You Know: More than 20 of Turkey's mountains are higher than 10,000 feet.

Ukraine

✦Capital: Kiev
✦Population: 47,425,336
✦Area: 233,090 sq. mi. (603,700 sq. km.)
✦Currency: $1 = 5.31 hryvnia
✦Language: Ukrainian, Russian
✦Did You Know: In the 1840s, Russian rulers banned the Ukrainian language from schools here.

Turkmenistan

✦Capital: Ashgabat
✦Population: 4,952,081
✦Area: 188,456 sq. mi. (488,100 sq. km.)
✦Currency: $1 = 5,200 manats
✦Language: Turkmen, Russian, Uzbek
✦Did You Know: Turkmen are famed for the beautiful carpets they weave from sheep wool.

United Arab Emirates

✦Capital: Abu-Dhabi
✦Population: 2,563,212
✦Area: 32,000 sq. mi. (82,880 sq. km.)
✦Currency: $1 = 3.67 dirhams
✦Language: Arabic, Persian, English, Hindi
✦Did You Know: Abu Dhabi is the largest of these seven states, occupying about 90% of the country's land area.

Tuvalu

✦Capital: Funafuti-Atoll
✦Population: 11,636
✦Area: 10 sq. mi. (26 sq. km.)
✦Currency: $1 = 1.31 Tuvalu dollars
✦Language: Tuvaluan, English
✦Did You Know: These low-lying islands are threatened by rising sea levels.

United Kingdom (Great Britain)

✦Capital: London
✦Population: 60,441,457
✦Area: 94,525 sq. mi. (244,820 sq. km.)
✦Currency: $1 = .54 pounds
✦Language: English
✦Did You Know: England, Northern Ireland, Scotland, and Wales make up the United Kingdom.

***All populations listed are 2005 estimates.**

Facts About Nations *(cont.)*

Tunisia	
Turkey	
Turkmenistan	
Tuvalu	
Uganda	
Ukraine	
United Arab Emirates	
Great Britain	

Use the information from Facts About Nations, page 48, to discover the number of languages spoken in each country. For each language, draw one book next to the name of the country.

Use the picture graph above to answer the following questions:

1. Which country has (or countries have) the most languages spoken? ____________________
2. Which country only has one official language? ____________________
3. How many languages are spoken in Tuvalu and Turkmenistan together? ____________________
4. If you were to visit the Ukraine, how many languages would you hear? ____________________
5. How many more languages are spoken in Uganda than Tunisia? ____________________

Challenge: Create a picture graph of the languages spoken by the children in your classroom. If everyone speaks only English, make a list of five languages besides English and have the children choose one language they would like to learn. Graph your findings with a picture graph. Choose a symbol to represent languages. Write five questions and answers to go with your findings.

November Birthdays

NOVEMBER

1 Lyle Lovett, singer/actor, 1957
2 Nelly, rapper, 1974
3 Roseanne, actress, 1952
4 Laura Bush, first lady, 1946
5 Ryan Adams, musician, 1974
6 James Naismith, basketball inventor, 1861
7 Marie Curie, scientist, 1867
8 Courtney Thorne-Smith, actress, 1967
9 Nick Lachey, singer, 1973
10 Eve, rapper/actress, 1978
11 Leonardo DiCaprio, actor, 1974
12 Anne Hathaway, actress, 1982
13 Robert Louis Stevenson, author, 1850
14 Condoleezza Rice, secretary of state, 1954
15 Zena Grey, actress, 1988
16 Marg Helgenberger, actress, 1958
17 Tom Seaver, baseball player, 1944
18 Owen Wilson, actor, 1968
19 Gail Devers, Olympic champion, 1966
20 Ming-Na Wen, actress, 1967
21 Ken Griffey Jr., baseball player, 1969
22 Jamie Lee Curtis, actress, 1958
23 Billy the Kid, outlaw, 1859
24 Scott Joplin, composer, 1868
25 Jenna and Barbara Bush, President Bush's daughters, 1981
26 Charles Schulz, cartoonist, 1922
27 Jimi Hendrix, musician, 1942
28 Jon Stewart, TV host, 1962
29 Louisa May Alcott, author, 1832
30 Ben Stiller, actor, 1965

November Birthdays *(cont.)*

Authors	
Music Careers	
TV Hosts/Actors	
Sports Careers	
Inventors/Scientists	
Outlaws	
Government/Politics	

Use the information from November Birthdays, page 50, to make a picture graph. For each of the categories above, count the number of person/s that will fit. Draw one hat for each person represented. If the person fits into two categories, choose the category the famous person is best known for.

Use the picture graph above to assist you in answering the following questions.

1. Which career celebrates the most birthdays in November? ______________________________
2. In which category would you put the two daughters of President Bush?____________________
3. How many musicians have their birthdays in November? ______________________________
4. How many more TV hosts and actors celebrate birthdays than inventors or scientists during the month of November?

   ____
5. What is the name of the outlaw who had a birthday in November?______________________

Challenge: Create your own picture graph using the information from page 50. Create your own categories, collect the information, and pick a symbol to draw to represent the information. Write five questions and answers for the picture graph.

Golf

Golf began in Scotland as early as the 1400s. The first golf course in the U.S. opened in 1888 in Yonkers, NY. The sport has grown to include both men's and women's professional tours. And millions play golf just for fun.

The men's tour in the U.S. is run by the Professional Golf Association (PGA). The four major championships (with the year first played) are:

British Open (1860)

United States Open (1895)

PGA Championship (1916)

Masters Tournament (1934)

The women's tour in the U.S. is guided by the Ladies Professional Golf Association (LPGA). The four major championships are:

United States Women's Open (1946)

McDonalds LPGA Championship (1955)

Nabisco Championship (1972)

Women's British Open (1976)

The All-Time "Major" Players

Here is a list of the pro golfers who've won the most major championships.

Men

1. Jack Nicklaus, 18
2. Walter Hagan, 11
3. Ben Hogan, 9
   Gary Player, 9
   Tiger Woods, 9
6. Tom Watson, 8

Women

1. Patty Berg, 15
2. Mickey Wright, 13
3. Louise Suggs, 11
4. Babe Didrikson Zaharias, 10
5. Betsy Rawls, 8
   Annika Sorenstam, 8

Golf *(cont.)*

Patty Berg	
Walter Hagen	
Ben Hogan	
Jack Nicklaus	
Gary Player	
Betsy Rawls	
Annika Sorenson	
Louise Suggs	
Tom Watson	
Tiger Woods	
Mickey Wright	

Use the information from Golf, page 52, to complete the picture graph. For each championship the golfer won, draw one cup.

Use the picture graph above to answer the following questions.

1. Which woman golfer won the most championships? ________________________________

2. How many more championships did the top man golfer win than the top woman? ____________

3. Which two woman golfers tied with the same amount of wins? ________________________

4. How many more championships did Tiger Woods win than Tom Watson? _______________

5. How many championships did all the woman golfers win combined? __________________

Challenge: Choose four different types of gum. Ask 20 friends which flavor of gum would be their favorite using the list of four kinds you have chosen. Create a picture graph along with five questions and answers to show the results of your information.

Native North Americans

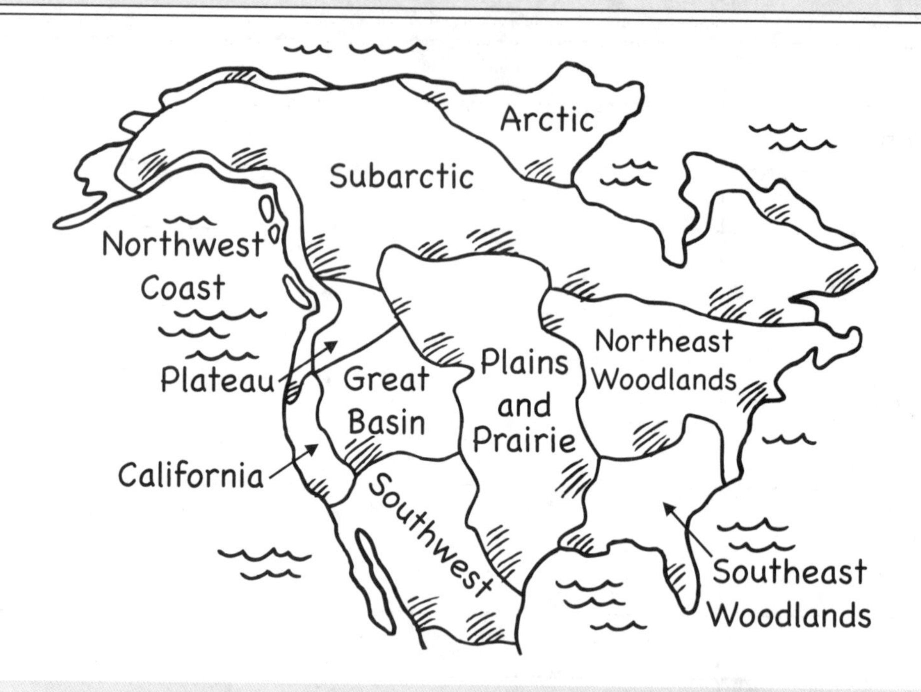

MAJOR CULTURAL AREAS OF NATIVE NORTH AMERICANS

Climate and geography influenced the culture of the people who lived in these regions. On the plains, for example, people depended on the great herds of buffalo for food. For Aleuts and Eskimos in the far North, seals and whales were an important food source. There are more than 560 tribes officially recognized by the U.S. government today and more than 56 million acres of tribal lands. Below are just a few well-known tribal groups that have lived in these areas.

NORTHEAST WOODLANDS
The Illinois, Iroquois (Mohawk, Onondaga, Cayuga, Oneida, Seneca, and Tuscarora), Lenape, Menominee, Micmac, Narragansett, Potawatomi, Shawnee.

SOUTHEAST WOODLANDS
The Cherokee, Chickasaw, Choctaw, Creek, Seminole.

PLAINS & PRAIRIE The Arapaho, Blackfoot, Cheyenne, Comanche, Hidatsa, Mandan, Sioux.

SOUTHWEST The Navajo, Apache, Havasupai, Mojave, Pima, Pueblo (Hopi, Isleta, Laguna, Zuñi).

GREAT BASIN The Paiute, Shoshoni, Ute.

CALIFORNIA The Klamath, Maidu, Miwok, Modoc, Patwin, Pomo, Wintun, Yurok.

PLATEAU The Cayuse, Nez Percé, Okanagon, Salish, Spokan, Umatilla, Walla Walla, Yakima.

NORTHWEST COAST The Chinook, Haida, Kwakiutl, Makah, Nootka, Salish, Tlinigit, Tsimshian, Tillamook.

SUBARCTIC The Beaver, Cree, Chipewyan, Chippewa, Ingalik, Kaska, Kutchin, Montagnais, Naskapi, Tanana.

ARCTIC The Eskimo (Inuit and Yipuk), Aleut.

Native North Americans *(cont.)*

Arctic	
California	
Great Basin	
Northeast Woodlands	
Northwest Coast	
Plains and Prairie	
Plateau	
Southeast Woodlands	
Southwest	
Subarctic	

Use the information on Major Cultural Areas of the Native North Americans, page 54, to complete the picture graph about Native Americans groups. (Count only the main groups—not the groups in parentheses.) To represent each tribal group in the regions, draw one triangle.

Use the picture graph above to answer the following questions.

1. Which area has the largest number of Native American Indian groups? ____________________

2. How area has the smallest number of groups? ____________________

3. How many groups are found in California and the Southwest combined?____________________

4. How many more are on the Plateau than on the Plains?____________________

5. How many Native American groups are there combined? ____________________

Challenge: Count the number of teachers in each grade in your elementary school. Create a picture graph to show how many teachers teach in each grade. Create the symbol of your choice to represent one teacher on the graph. Write five questions and answers for your graph.

Body Basics

Your body is made up of many different parts that work together every minute of every day and night. It's more amazing than any machine or computer. Even though everyone's body looks different outside, people have the same parts inside. Each system of the body has its own job. Some of the systems also work together to keep you healthy and strong.

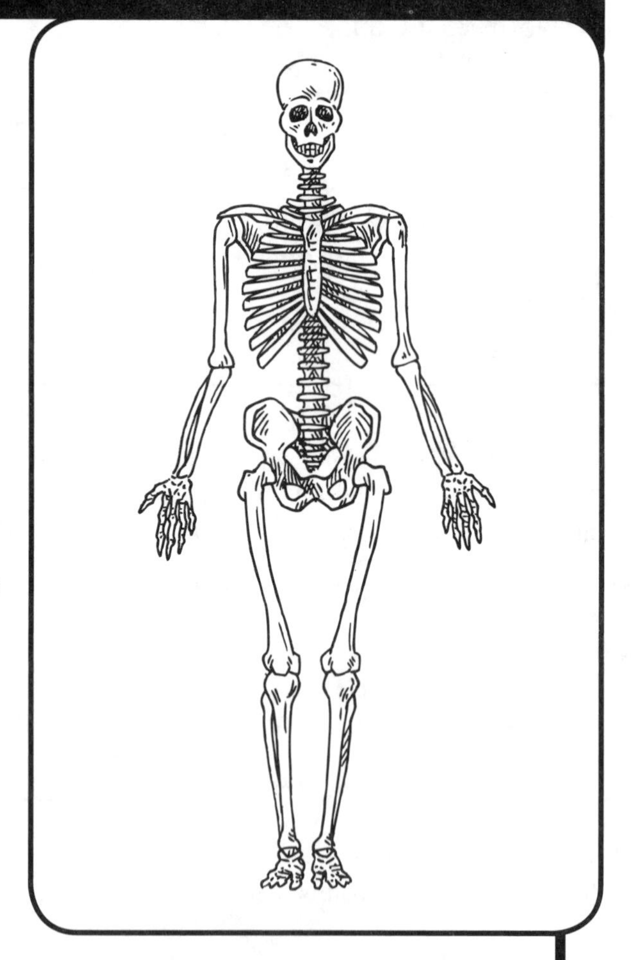

CIRCULATORY SYSTEM In the circulatory system, the heart pumps blood, which then travels through tubes, called arteries, to all parts of the body. The blood carries the oxygen and food that the body needs to stay alive. Veins carry the blood back to the heart.

DIGESTIVE SYSTEM The digestive system moves food through parts of the body called the esophagus, stomach, and intestines. As the food passes through, some of it is broken down into tiny particles called nutrients, which the body needs. Nutrients enter the bloodstream, which carries them to all parts of the body. The digestive system then changes the remaining food into waste that is eliminated from the body.

ENDOCRINE SYSTEM
The endocrine system includes glands that are needed for some body functions. There are two kinds of glands. Exocrine glands produce liquids such as sweat and saliva. Endocrine glands produce chemicals called hormones. Hormones control body functions, such as growth.

NERVOUS SYSTEM The nervous system enables us to think, feel, move, hear, and see. It includes the brain, the spinal cord, and nerves in all parts of the body. Nerves in the spinal cord carry signals back and forth between the brain and the rest of the body. The brain tells us what to do and how to respond. It has three major parts. The cerebrum controls thinking, speech, and vision. The cerebellum is responsible for physical coordination. The brain stem controls the respiratory, circulatory, and digestive systems.

RESPIRATORY SYSTEM The respiratory system allows us to breathe. Air comes into the body through the nose and mouth. It goes through the windpipe (or trachea) to two tubes (called bronchi), which carry air to the lungs. Oxygen from the air is taken in by tiny blood vessels in the lungs. The blood then carries oxygen to the cells of the body.

Body Basics *(cont.)*

respiratory	
endocrine	
digestive	
circulatory	
nervous	

Use the information from Body Basics, page 56, to complete the picture graph. For each item in the system, draw one circle.

Use the picture graph above to answer the following questions.

1. Which system has the most parts in it? ________________________
2. Which system has the least? ________________________
3. How many more parts are in the nervous system than in the respiratory? ____________
4. How many parts are in the nervous and the endocrine combined? ____________
5. How many parts make up all five systems? ________________________

Challenge: Take a survey in your classroom of 10 children. Record the number of cousins they have. Create a picture graph to show the information you have collected. Create the picture/object of your choice to represent cousins. Write five questions and answers to go with the information you discovered.

The Planets

1 MERCURY

Average distance from the Sun: 36 million miles
Diameter: 3,032 miles
Average temp.: 333° F
Surface: silicate rock
Time to revolve around the Sun: 88 days
Day (synodic—midday to midday): 175.94 days
Number of moons: 0

DID YOU KNOW? Mercury is the closest planet to the Sun, but it gets very cold there. Since Mercury has almost no atmosphere, most of its heat escapes at night, and temperatures can fall to –300°.

2 VENUS

Average distance from the Sun: 67 million miles
Diameter: 7,521 miles
Average temp.: 867° F
Surface: silicate rock
Time to revolve around the Sun: 224.7 days
Day (synodic—midday to midday): 116.75 days
Number of moons: 0

DID YOU KNOW? Venus rotates in the opposite direction from all the other planets. Unlike on Earth, on Venus the sun rises in the west and sets in the east.

3 EARTH

Average distance from the Sun: 93 million miles
Diameter: 7,926 miles
Average temp.: 59° F
Surface: water, basalt and granite rock
Time to revolve around the Sun: 365 1/4 days
Day (synodic—midday to midday): 24h
Number of moons: 1

DID YOU KNOW? The Earth travels around the Sun at a speed of more than 66,000 miles per hour.

4 MARS

Average distance from the Sun: 142 million miles
Diameter: 4,213 miles
Average temp.: –81° F
Surface: iron-rich basaltic rock
Time to revolve around the Sun: 687 days
Day (synodic—midday to midday): 24h 39m 35s
Number of moons: 2

DID YOU KNOW? In 1877, astronomer Giovanni Schiaparelli thought he saw lines on Mars, which he called "channels," or "canali" in Italian. This was mistranslated into English as "canals," making people think there were canal-building Martians.

5 JUPITER

Average distance from the Sun: 484 million miles
Diameter: 88,732 miles
Average temp.: –162° F
Surface: liquid hydrogen
Time to revolve around the Sun: 11.9 years
Time to rotate on its axis: 9h, 55m, 30s
Number of moons: 63

DID YOU KNOW? The 4 largest moons were discovered by Galileo in 1610; 21 others were not found until 2003.

6 SATURN

Average distance from the Sun: 887 million miles
Diameter: 74,975 miles
Average temp.: –218° F
Surface: liquid hydrogen
Time to revolve around the Sun: 29.5 years
Day (synodic—midday to midday): 10h 39m 23s
Number of moons: 47

DID YOU KNOW? Using a simple early telescope, Galileo discovered what turned out to be rings around Saturn.

7 URANUS

Average distance from the Sun: 1.8 billion miles
Diameter: 31,763 miles
Average temp.: –323° F
Surface: liquid hydrogen and helium
Time to revolve around the Sun: 84 years
Day (synodic—midday to midday): 17h 14m 23s
Number of moons: 27

DID YOU KNOW? Because Uranus is tipped 98 degrees on its axis, its seasons are far more extreme than those of Earth: the north pole is dark for 42 years at a time.

8 NEPTUNE

Average distance from the Sun: 2.8 billion miles
Diameter: 30,603 miles
Average temp.: –330° F
Surface: liquid hydrogen and helium
Time to revolve around the Sun: 164.8 years
Day (synodic—midday to midday): 16d 6h 37m
Number of moons: 13

DID YOU KNOW? Neptune was discovered in 1846, after British astronomer John Adams and French mathematician Urbain Le Verrier independently predicted where it would be, based on its effect on Uranus's orbit.

9 PLUTO

Average distance from the Sun: 3.6 billion miles
Diameter: 1,485 miles
Average temp.: –369° F
Surface: rock and frozen gases
Time to revolve around the Sun: 247.7 years
Day (synodic—midday to midday): 6d 9h 17m
Number of moons: 3

DID YOU KNOW? Pluto is the smallest planet. Some scientists do not consider it a planet, just one of many large objects orbiting the Sun outside Neptune's orbit.

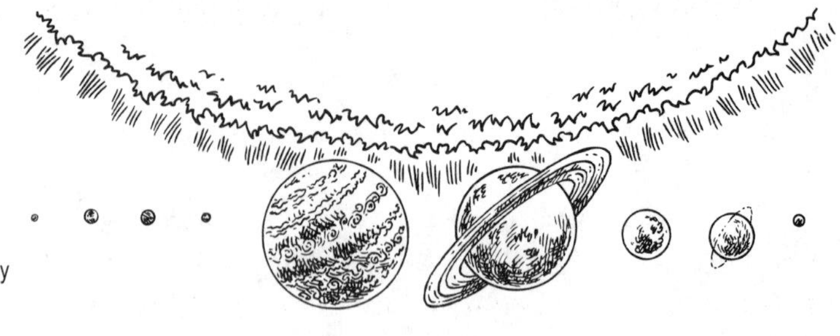

The Planets *(cont.)*

Earth	Juniper	Mars	Mercury	Neptune	Pluto	Saturn	Uranus	Venus

Use the information from The Planets, page 58, to record the number of moons each planet has.

List the number of moons in order from least to greatest.

_______, _______, _______, _______, _______, _______, _______, _______, _______

Make a stem and leaf plot from the data above. Create a title and a key for your leaf and stem plot.

____

(Title)

Stem	Leaves

Key: _______________ _______________ = _______________ moons on planets

Use the stem and leaf plot to answer the following questions:

1. What is the lowest number of moons displayed? _______________
2. What is the highest number of moons? _______________
3. Are the numbers of moons spread out evenly or clustered? _______________
4. What observations can we make from this stem and leaf plot? _______________

Challenge: Take a survey of 10 adults you know. Ask them to tell you how old they are. Use this information to create a stem and leaf plot. List the names of the adults and the ages gathered. Write four questions and answers to accompany the stem and leaf plot.

Comparing Money

Kenya

- ✦Capital: Nairobi
- ✦Population: 33,829,590*
- ✦Area: 224,962 sq. mi. (582,650 sq. km.)
- ✦Currency: $1 = 77.19 shillings
- ✦Language: Swahili, English
- ✦Did You Know: The coast of this east African nation has seen the influence of traders from Persia, China, the Malay peninsula, Portugal, and England.

Kyrgyzstan

- ✦Capital: Bishkek
- ✦Population: 5,146,281
- ✦Area: 76,641 sq. mi. (198,500 sq. km.)
- ✦Currency: $1 = 41.2 soms
- ✦Language: Kyrgyz, Russian
- ✦Did You Know: This Central Asian country is almost entirely mountainous.

Luxembourg

- ✦Capital: Luxembourg
- ✦Population: 468,571
- ✦Area: 998 sq. mi. (2,586 sq. km.)
- ✦Currency: $1 = 77 euros
- ✦Language: French, German
- ✦Did You Know: This country takes its name from a Roman castle on the Alzette River whose name, Lucilinburhuc, meant "Little Fortress."

Mauritius

- ✦Capital: Port Louis
- ✦Population: 1,230,602
- ✦Area: 788 sq. mi. (2,040 sq. km.)
- ✦Currency: $1 = 28.57 Mauritian rupees
- ✦Language: English, French, Creole, Hindi
- ✦Did You Know: The dodo became extinct here by 1681, 83 years after the Dutch arrived.

Mexico

- ✦Capital: Mexico City
- ✦Population: 106,202,903
- ✦Area: 761,606 sq. mi. (1,972,550 sq. km.)
- ✦Currency: $1 = 11.24 pesos
- ✦Language: Spanish, Mayan dialects
- ✦Did You Know: The Aztec capital of Tenochtitlan was destroyed by the Spanish in 1521.

Moldova

- ✦Capital: Chisinau
- ✦Population: 4,455,421
- ✦Area: 13,067 sq. mi. (33,843 sq. km.)
- ✦Currency: $1 = 12.52 lei
- ✦Language: Moldovan, Russian
- ✦Did You Know: Grapes are a major crop, in Moldova, and winemaking is a major industry.

Nepal

- ✦Capital: Kathmandu
- ✦Population: 27,676,547
- ✦Area: 54,363 sq. mi. (140,800 sq. km.)
- ✦Currency: $1 = 70.95 rupees
- ✦Language: Nepali, many dialects
- ✦Did You Know: Mt. Everest, the world's highest mountain, is partly in Nepal.

Nicaragua

- ✦Capital: Managua
- ✦Population: 5,465,100
- ✦Area: 49,998 sq. mi. (129,494 sq. km.)
- ✦Currency: $1 = 16.18 gold córdobas
- ✦Language: Spanish
- ✦Did You Know: The eastern shore is called Costa de Mosquitos (Mosquito Coast).

***All populations listed are 2005 estimates.**

Comparing Money *(cont.)*

Luxembourg	Moldova	Mauritius	Mexico	Nepal	Nicaragua	Kenya	Kyrgyzstan

Use the information from Comparing Money, page 60; record the information comparing the money of that country to the dollar bill. Round the money of each country to the nearest whole number and record it on the table.

List the currency exchange in order from least to greatest

________, ________, ________, ________, ________, ________, ________, ________

Make a stem and leaf plot from the data above. Create a title and a key for your leaf and stem plot.

____

(Title)

Stem	Leaves

Key: ________________ ________________ = ______________________________ currency

Use the stem and leaf plots to answer the following questions:

1. What is the country with the currency furthest in value from the dollar? ________________
2. Which country has the currency, which is closest in value to the dollar?________________
3. Are the currency rates spread out evenly or clustered? ________________
4. What observations can we make from this stem and leaf plot? ________________

Challenge: Find out the temperatures in 10 cities on the same day. Create a stem and leaf plot to show the information you gathered. Record the names and temperatures of the cities. Write four questions and answers to accompany your stem and leaf plot.

Basketball

Basketball began in 1891 in Springfield, Massachusetts, when Dr. James Naismith invented it, using peach baskets as hoops. At first, each team had nine players instead of five. Big-time pro basketball started in 1949, when the National Basketball Association (NBA) was formed. The Women's National Basketball Association (WNBA) began play in 1997.

HIGHLIGHTS OF THE 2004–2005 NBA SEASON

SCORING LEADER:
Allen Iverson,
Philadelphia 76ers

Games: 75
Points: 2,302
Average: 30.7

REBOUNDING LEADER:
Kevin Garnett,
Minnesota Timberwolves

Games: 82
Rebounds: 1,108
Average: 13.5

BLOCKED SHOTS LEADER:
Andrei Kirilenko, Utah Jazz

Games: 41
Blocks: 136
Average: 3.32

STEALS LEADER:
Larry Hughes,
Washington Wizards

Games: 61
Steals: 176
Average: 2.89

ASSISTS LEADER:
Steve Nash,
Phoenix Suns

Games: 75
Assists: 861
Average: 11.5

HALL OF FAME

The Naismith Memorial Hall of Fame in Springfield, Massachusetts, was founded to honor great basketball players, coaches, referees, and others important to the history of the game. The newest class, heading for the hall in September 2005, includes college coaching legends Jim Calhoun and Jim Boeheim, NBA coach and broadcaster Hubie Brown, and women's coach Sue Gunter. International star Hortencia de Fatima Marcari of Brazil, who led her country to a World Championship in 1994, will also be enshrined.
www.hoophall.com

Prior to the 2004-2005 season, the Chicago Bulls had not made the NBA playoffs since Michael Jordan led them to a sixth championship in the 1997-1998 season. In that span, the Bulls lost 341 games. This year they started the season 0-9, but clinched a playoff spot before the season ended.

Basketball *(cont.)*

Alan Iverson	Andrei Kirilenko	Kevin Garnett	Larry Hughes	Steve Nash

Use the information from Basketball, page 62, to record the number of games played by the players listed on the table.

List the games played in order from least to greatest

________, ________, ________, ________, ________

Make a stem and leaf plot from the data above. Create a title and a key for your stem and leaf plot.

____

(Title)

Stem	Leaves

Key: ______________ ______________ = ______________________ games played

1. What is the lowest number of games played? ______________________
2. What is the highest number of games played? ______________________
3. Are the numbers of games spread out evenly or clustered? ______________________
4. What observations can we make from this stem and leaf plot? ______________________

Challenge: Ask 10 friends to pick their favorite number between 11 and 99. Record the information and create a stem and leaf plot to show what you discovered. Record the name of the friends and their favorite number. Write four questions and answers to accompany the stem and leaf plot.

Unique Holidays

OTHER SPECIAL HOLIDAYS

Valentine's Day—February 14 is a day for sending cards or gifts to people you love.
Mother's Day and **Father's Day**—Mothers are honored on the second Sunday in May. Fathers are honored on the third Sunday in June.
Grandparents' Day—This day to honor grandparents comes every year on the first Sunday after Labor Day.
Halloween—In ancient Britain, Druids wore grotesque costumes on October 31 to scare off evil spirits. Today, while "trick or treating," children ask for candy or money for UNICEF, the United Nations Children's Fund.
Hanukkah (also Chanukah)—This eight-day Jewish festival of lights began on the evening of December 25 in 2005 and December 15 in 2006.
Kwanzaa—This seven-day African-American festival begins on December 26. It celebrates seven virtues: unity, self-determination, collective work and responsibility, cooperative economics, purpose, creativity, and faith.

ODD HOLIDAYS

You can chase away the "back-to-school blues" by observing Hobbit Day or Elephant Appreciation Day on September 22. Here are a few other odd "days" you've probably never heard of:

January
7: I'm Not Going to Take It Anymore Day
18: Pooh Day (A.A. Milne's Birthday)
21: Squirrel Appreciation Day

February
14: Ferris Wheel Day
15: National Gum Drop Day

March
1: National Pig Day
18: Awkward Moments Day
22: International Goof-Off Day
25: Pecan Day

April
10: National Siblings Day
11: Barbershop Quartet Day
30: National Honesty Day

May
1: Save the Rhino Day
12: Limerick Day
16: International Sea Monkey Day

June
2: National Bubba Day
18: World Juggling Day
28: National Handshake Day

July
11: International Town Criers Day
17: National Ice Cream Day
28: National Drive-Thru Day

August
6: National Fresh Breath Day
13: National Underwear Day
17: Sandcastle Day

September
4: Newspaper Carrier Day
12: Video Games Day
19: Talk Like a Pirate Day

October
1: Scare a Friend Day
26: Mule Day
31: National Knock-Knock Day

November
6: Saxophone Day
21: World Hello Day
25: Buy Nothing Day

December
4: National Dice Day
16: Underdog Day
31: Make Up Your Mind Day

Special note: September 1–30 is Be Kind to Writers and Editors Month!

Unique Holidays *(cont.)*

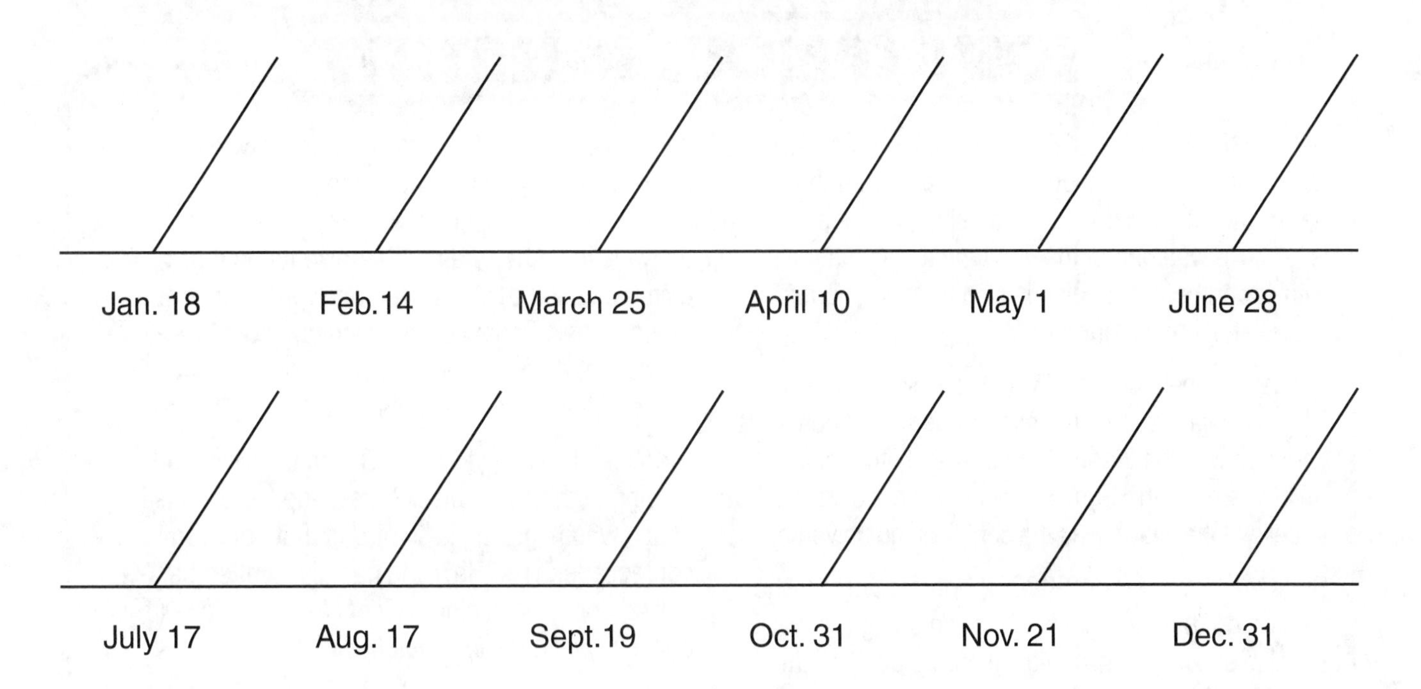

Use the information from Unique Holidays, page 64, to write the names for the odd holidays which fit the dates which match the time line.

Answer the following questions using the timeline above:

1. On what day would you celebrate a ride at the fair?____________________________
2. What date would you celebrate with a handshake? ____________________________
3. Where would be the best place to be on August 17? ____________________________
4. On July 17, why is it important to go to the store for something cold? ________________
5. On October 31, what kind of things will you tell your friends?____________________

Challenge: Create your own time line using at least 10 odd holidays that have not been used from page 64. Write five questions and answers to accompany your time line.

Famous Scientists

Some Famous Scientists

ARCHIMEDES (about 287 b.c.–212 b.c.), Greek mathematician and inventor who discovered that heavy objects could be moved using pulleys and levers. He was one of the first to test his ideas with experiments. He also is said to have shouted "Eureka!" ("I have found it!").

Nicolaus Copernicus (1473–1543), Polish scientist who is known as the founder of modern astronomy. He came up with the theory that Earth and other planets revolve around the Sun. But most thinkers continued to believe that Earth was the center of the universe.

Sir Isaac Newton (1642–1727), British scientist who worked out the basic laws of motion and gravity. He also showed that sunlight is made up of all the colors of the rainbow. He invented the branch of mathematics called calculus about the same time as the German scientist Gottfried von Leibniz (1646–1716), who was the first to make it widely known.

Charles Darwin (1809–1882), British scientist who is best known for his theory of evolution by natural selection. According to this theory, living creatures, by gradually changing so as to have the best chances of survival, slowly developed over millions of years into the forms they have today.

George Washington Carver (1864–1943), born in Missouri of slave parents, became world-famous for his agricultural research. He found many nutritious uses for peanuts and sweet potatoes, and taught farmers in the South to rotate their crops to increase their yield.

Albert Einstein (1879–1955), German-American physicist who developed revolutionary theories about the relationships between time, space, matter, and energy. He won a Nobel Prize in 1921.

Rachel Carson (1907–1964), U.S. biologist and leading environmentalist whose 1962 book Silent Spring warned that chemicals used to kill pests were killing harmless wildlife. Eventually DDT and certain other pesticides were banned in the U.S.

Jane Goodall (1934–), British scientist who is a leading authority on chimpanzee behavior. Goodall discovered that chimpanzees use tools, such as twigs to "fish" for ants. She also found that chimpanzees have complex family structures and personalities. Today, Goodall writes books, creates movies, and speaks publicly as an advocate for the preservation of wild habitats.

Stephen Hawking (1942–), British physicist and leading authority on black holes—dense objects in space whose gravity is so strong that not even light can escape them. Hawking has also written best-selling books, including *A Brief History of Time* (1988) and *The Universe in a Nutshell* (2001).

Linda Spilker (1955–), space scientist who is deputy project scientist for the current Cassini mission to Saturn. The Cassini orbiter is expected to orbit Saturn for several years, measuring and recording data on Saturn, its rings, and its 47 known moons. "Saturn's rings have always fascinated me," Spilker says. "Now I can bring some of the new ring data back to earth."

Paul Sereno (1957–), American paleontologist who has traveled over much of the world to discover and study early dinosaur fossils. His research has helped explain dinosaur evolution and behavior.

Famous Scientists *(cont.)*

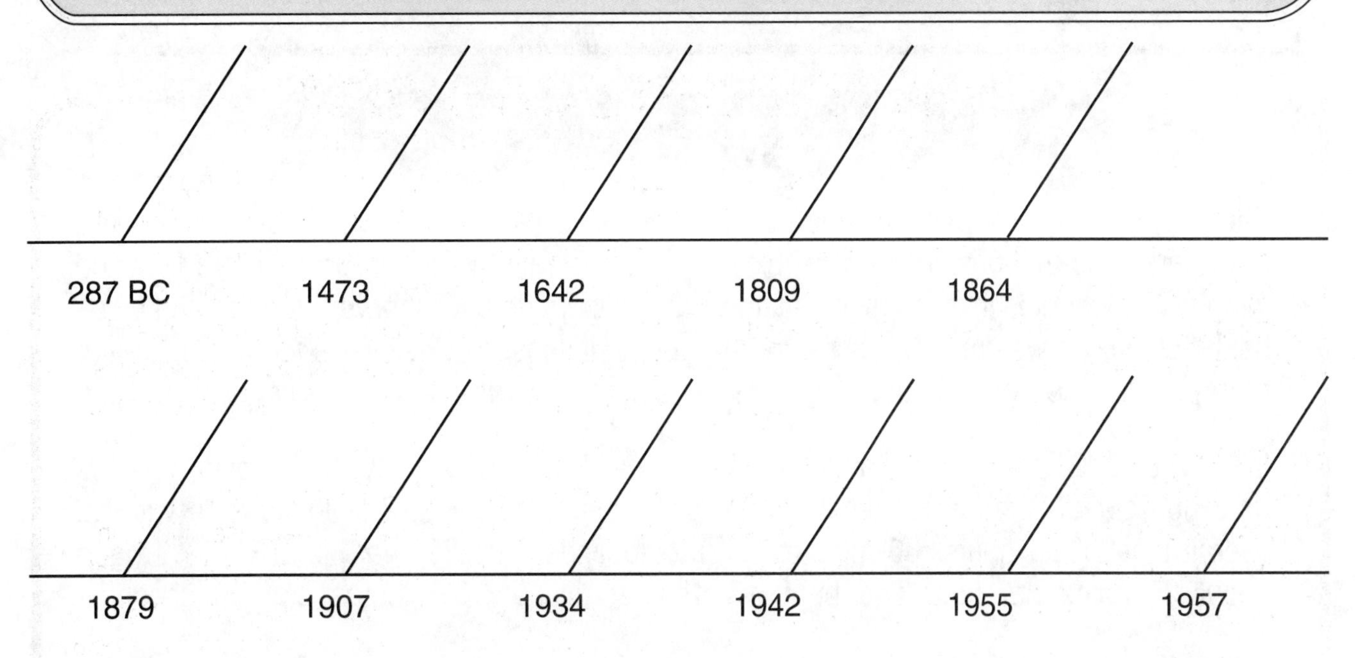

Use the information from Famous Scientists, page 66, to record the names of the famous scientists whose birth years match the dates on the time line. Write the name of each famous scientist above the year he/she was born.

Answer the following questions using the time line above.

1. Who was born in 287 BC? ______________________________

2. Which two scientists were born only two years apart? ______________________________

3. Which scientists were born in the 1800s? ______________________________

4. How many scientists were born before Christ? ______________________________

5. Which scientists were born in the 1900s? ______________________________

Challenge: Create your own time line using showing the birthdates of at least five people in your family. Write five questions and answers to go with your time line.

TOUR DE LANCE!

The "modern" bicycle, with two wheels the same size, pedals, and a chain drive, appeared at the end of the 1800s. The name "bicycle" itself was first used in 1869. Before that, the various two-wheeled inventions were known as "velocipedes." Clubs were formed and races held, but cycling was mostly a sport for the upper classes; bicycles were too expensive for most people. The world's best-known cycling race, the Tour de France, was first held in 1903.

In July of 2004, American Lance Armstrong became the only cyclist to ever win the grueling Tour de France six times. He finished the three-week, 2,110-mile race with an overall time of 83 hours, 36 minutes, and 2 seconds. The 32-year-old Texan averaged about 25 miles per hour and beat his nearest rival by 6 minutes and 19 seconds.

Lance's first win, in 1999, was very special. In 1996, he had been diagnosed with cancer. He had two operations and went through chemotherapy. This didn't stop him. In May 1998, he came back and signed with the U.S. Postal Service Team, setting his sights on the Tour de France. In 2003, Lance tied the previous record of five Tour de France wins. Only four other cyclists—Jacques Anquetil, Bernard Hinault, Miguel Indurain, and Eddy Merckx—had ever achieved that feat.

Lance was born September 18, 1971, and raised by his mother in Plano, Texas. When he was 13, he won the first Iron Kids Triathlon (1985), beating lots of bigger kids. The cycling part was his favorite. By 1991, he was the U.S. national amateur champion.

Cycling. . . Tour de Lance *(cont.)*

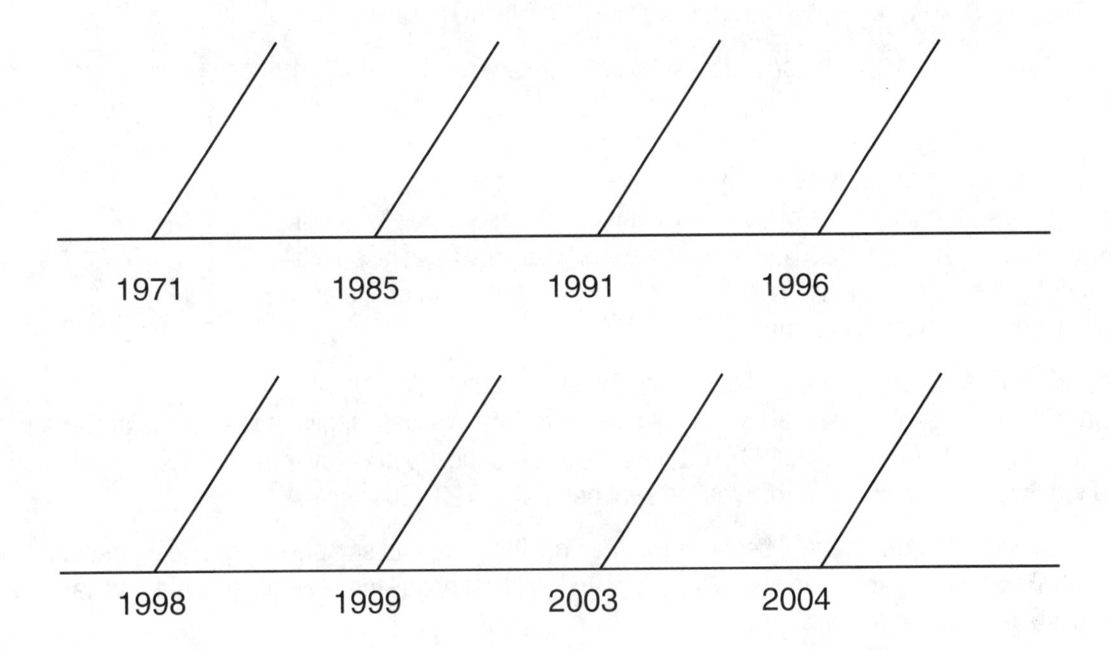

Use the information from Cycling. . . Tour De Lance!, page 68, to gather record events from the life of Lance Armstrong for the dates given on the timeline. Answer the following questions using the timeline above:

1. What year was Lance born? ______________________________
2. What year was he diagnosed with cancer? ______________________________
3. What year was it when Lance won the Kids Triathlon? ______________________________
4. In 1998, with what team did Lance sign? ______________________________
5. How old was Lance when he won the Tour de France for the 6th time? ______________

Challenge: Create your own time line using showing seven or more events in the life of one person you respect or admire. Write five questions and answers to accompany your time line.

Kid Inventors

K-K GREGORY, AGE 10, INVENTOR OF WRISTIES

Kathryn (K-K) Gregory from Bedford, Massachusetts, became an inventor in 1994 at the age of 10. She began experimenting with different ways to play outside in the snow and keep the snow from finding its way up the sleeve of her coat. With some help from her mom, K-K sewed synthetic fleece into cylinders that would fit under her sleeve but over her gloves. However, in field tests, snow was still sneaking under the lower edge. Then K-K altered her design so that it could be worn under gloves. This time her invention did the job and became an instant hit with other kids in her Girl Scout troop.

When K-K knew she had a winning idea, she contacted a patent attorney, who found out that her idea was original. She named her invention Wristies, applied for a patent, and even started a company to market it. In 1997 K-K became the youngest person ever to promote a product on the QVC TV network, where in just 6 minutes the spot generated $22,000 in sales.

K-K went a long way, step by step, beginning with an original idea and sample design, going through field testing and revision, market testing, patent search and application, all the way to starting a company, selling the product, and making a profit!

RICHIE STACHOWSKI, AGE 15, INVENTOR OF WATER TALKIES

When Richie Stachowski went snorkeling on his vacation, he wished he could shout out to tell everybody how beautiful the fish were. But he couldn't because he was underwater. After returning home, he researched underwater acoustics on the Internet and then began to build sample designs of an underwater megaphone. Using $267 from his savings account, he was able to produce Water Talkies, a toy that lets kids talk underwater. The prototype was made from a plastic cone, a snorkel mouthpiece, and a blow valve with a plastic membrane to keep the water out. Richie formed his own company and soon began selling his Water Talkies to major retailers.

2004 Craftsman/NSTA Young Inventors Award

Every year, the National Science Teachers Association (NSTA) and the Craftsman tool company hold a contest, judging the best new inventions created by kids. In 2004, these two students each won the top prize, a $10,000 U.S. savings bond.

Nicolette Mann, Christiansburg, Virginia, grade 4.
Invention: "Piano Peddles for Young Beginners." This box has three 5-inch wooden poles placed on piano pedals so that children can reach them. Nicolette designed them with her little brother in mind.

Katelyn Eubank, Indianola, Iowa, grade 7.
Invention: "The Easy Door Assist." This contraption puts four vertical rollers on the sides of a wheelchair. This helps older or disabled people get through doors more easily. Katelyn designed the invention to help her grandmother.

Kid Inventors

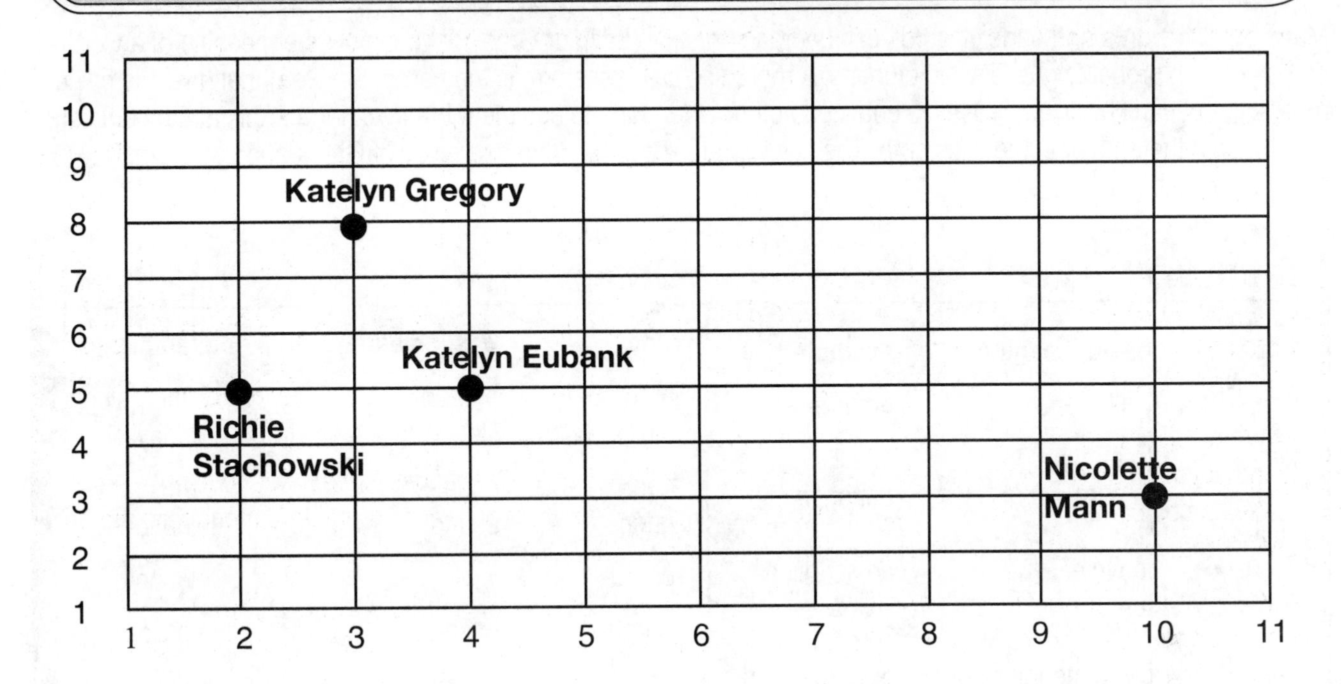

Name the ordered pair for each clue given below.

1. This is the person who invented something to keep snow out of your sleeves.

   ____

2. This is the person who made a toy for talking under water. ____________________________

3. This is the person who invented something to help small piano players reach the pedals.

   ____

4. This is the person who was thinking of her grandmother when she made these rollers

   ____

Graph and label the following points on the grid using the clues below.

5. The name of the invention that keeps snow out of your sleeves (5, 7)

6. The name of the invention used under water.(8, 2)

7. The name of the inventions to make playing piano easier (1, 5)

8. The name of the invention to help wheelchair patients get in and out of doors. (5, 2)

9. The name for someone who is not old, but has invented something (7, 7)

Getting to the Root

Many English words and parts of words can be traced back to Latin or Greek. If you know the meaning of a word's parts, you can probably guess what it means. A root (also called a stem) is the part of the word that gives its basic meaning, but can't be used by itself. Roots need other word parts to complete them: either a prefix at the beginning, or a suffix at the end, or sometimes both. The following tables give some examples of Greek and Latin roots, prefixes, and suffixes.

Latin

root	basic meaning	example
-gress-	walk	progress
-ject-	to throw	reject
-port-	to carry	transport
-scrib-/ -script-	to write	prescription
-vert-	turn	invert
prefix	basic meaning	example
de-	away, off	defrost
inter-	between, among	international
non-	not	nontoxic
pre-	before	prevent
re-	again, back	rewrite
trans-	across, through	trans-Atlantic
suffix	basic meaning	example
-ation	(makes verbs into nouns)	invitation
-fy/-ify	make or cause to become	horrify
-ly	like, to the extent of	highly
-ment	(makes verbs into nouns)	government
-ty/-ity	state of	purity

Greek

root	basic meaning	example
-anthrop-	human	anthropology
-bio-	life	biology
-dem-	people	democracy
-phon-	sound	telephone
-photo-	light	telephoto
-scope-	to see	telescope
prefix	basic meaning	example
anti-/ant-	against	antisocial
auto-	self	autopilot
biblio-/ bibl-	book	bibliography
micro-	small	microscope
tele-	far off	television
suffix	basic meaning	example
-graph	write, draw, describe, record	photograph
-ism	act, state, theory of	realism
-ist	one who believes in, practices	capitalist
-logue/ -log	speech, to speak	dialogue
-scope	see	telescope

Getting to the Root *(cont.)*

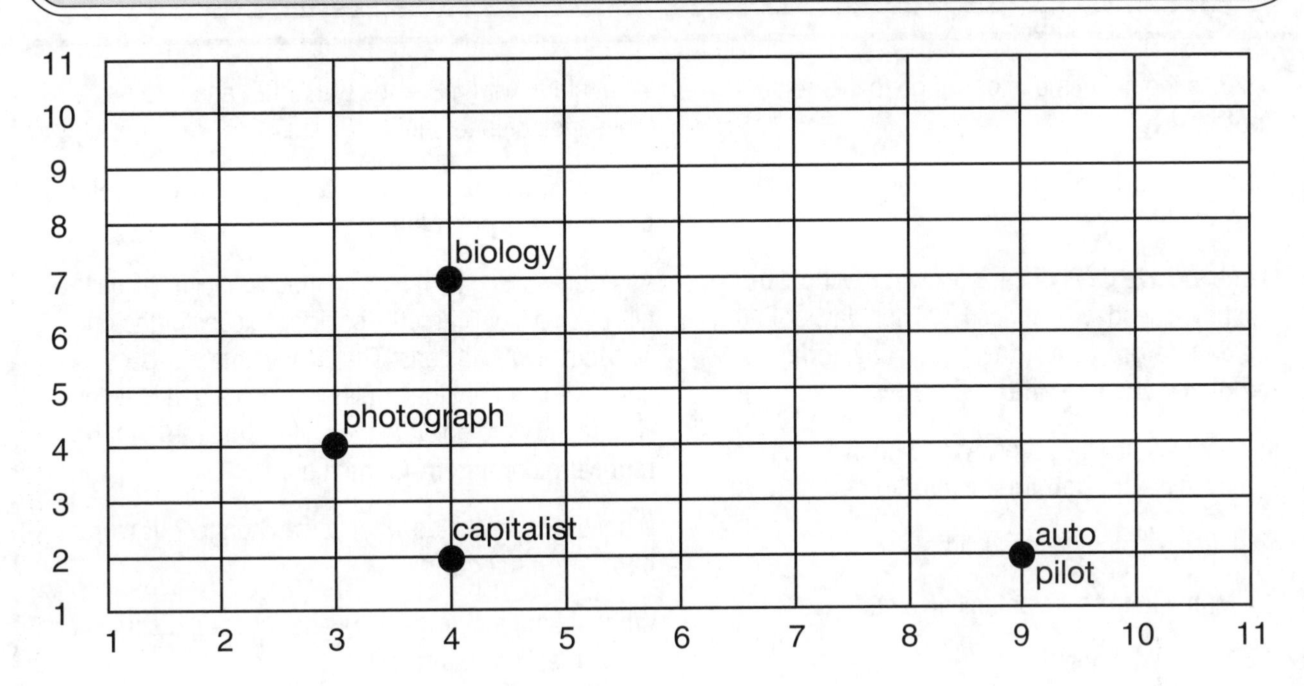

Name the ordered pair for each clue given below.

1. This is a person who makes a living by putting his or her money into other businesses or ideas that he or she is in charge of. ______________________________
2. This is a machine that runs a vehicle without a human's help. ______________________________
3. This is a picture that comes from a camera. ______________________________
4. This is a name for the science that studies living things. ______________________________

Graph and label the following points on the grid.

6. What do you call the instrument used to see objects in space? (6,7)
7. What do you call the instrument used to study very small animals, plants or bacteria? (2,2)
8. What do you call the study of human beings? (1,9)
9. In what kind of art from Europe did painters paint pictures that were like real life? (8,2)
10. What do you call a person who does not like to be around other people? (9,7)

Jokes and Riddles

When is the best time of day to go to the dentist? **Tooth-hurty**

A farmer had 12 cows. All but 9 died. How many cows did he have left? **9**

Name the five days of the work week without using Monday, Tuesday, Wednesday, Thursday, or Friday. **the day before yesterday, yesterday, today, tomorrow, the day after tomorrow**

In what way are the letter "A" and noon exactly the same? **They're both in the middle of day.**

Yo: I just got bit by a mosquito.

Bo: Man, I hate those arithmetic bugs!

Yo: Arithmetic bugs?

Bo: Sure. They add to misery, subtract from fun, divide your attention, and multiply quickly!

What newspaper did cave people read? **The Prehistoric Times**

Zip: What's worse than finding a worm in your apple?

Zap: I don't know, that sounds pretty bad.

Zip: Not as bad as finding half of a worm!

What did the boyfriend melon say to the girlfriend melon? **Cantaloupe tonight. Dad has the car.**

Dad: Where are you going with my toolbox?

Son: It's for my math homework.

Dad: How can tools help you with math?

Son: The instructions said to find multi-pliers.

Why didn't the skeleton cross the road? **She didn't have the guts**!

I'm the beginning of eternity, the end of space and time, the middle of every buzzing bee, and the end of every rhyme. What am I? **the letter "E"**

If 10 robins can catch 10 worms in 10 minutes, how long will it take one robin to catch a worm? **10 minutes**

What starts with a P, ends with an E, and has thousands of letters in it? **Post Office**

Why did Cinderella's soccer team always lose? **Her coach was a pumpkin.**

As I was walking to the mall, I met eight girls, all quite tall. Each tall girl carried a squirrel, except for the one whose hair was in curls. They also came with six young boys, who brought their mothers who carried their toys. How many were going to the mall? **Just me. All the rest were coming from the mall.**

What did the clock do when it was hungry? **It went back "four" seconds.**

What did the mover get when he dropped a computer on his toes? **Megahertz**

Here's that letter "A" again! How is it like a flower? **Because a "B" is always after it.**

The ancient "Riddle of the Sphinx":

From Greek mythology, this is the oldest known riddle. The heroic Oedipus knew the answer, do you?

What animal walks on four feet in the morning, two at noon, and three in the evening?

The answer is a man. He crawls on all fours as a baby, walks on two legs when grown, and uses a cane in old age.

Jokes and Riddles *(cont.)*

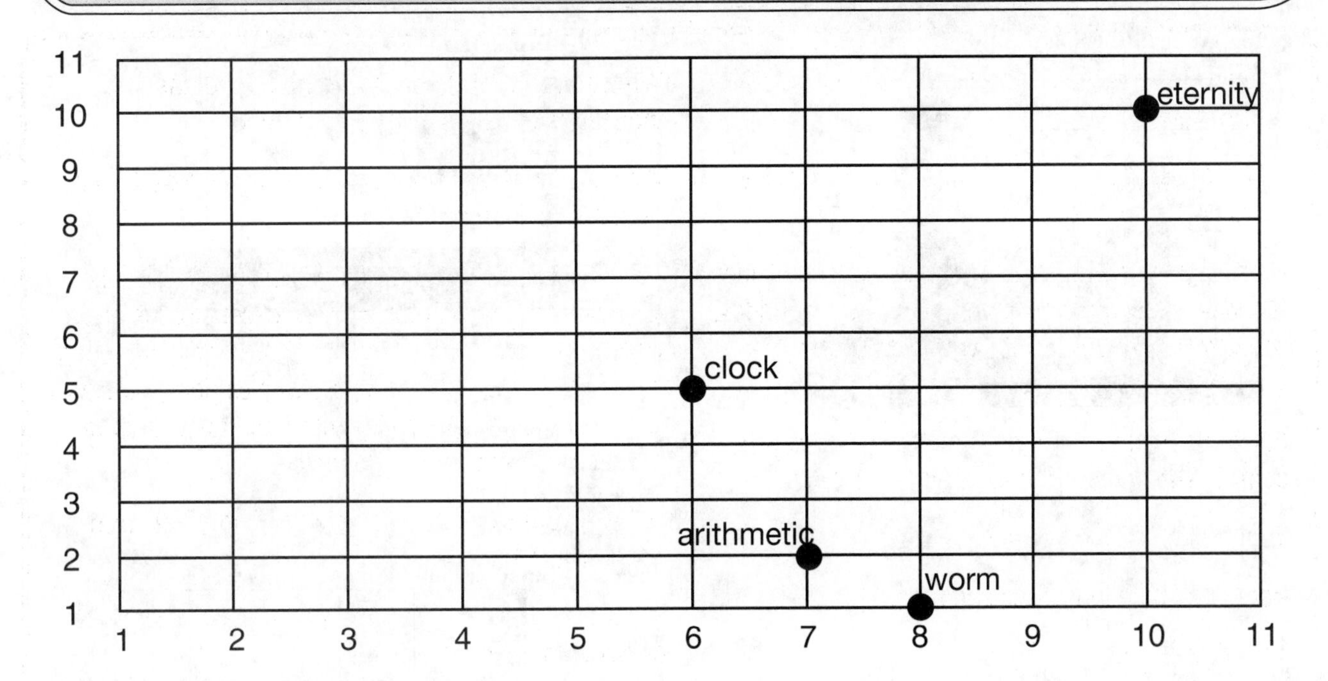

Name the ordered pair for each clue given below.

1. You use this to tell time. ______________________

2. This is a subject in school which uses numbers ______________________.

3. Robins eat this. ______________________

4. This word means goes on forever and ever. It has no end. ______________________

Graph and label the words which answer each riddle below.

5. The name for a person who runs a farm. (5,7)

6. This is a round fruit that grows in a patch. (4,2)

7. This is not a real word. It means an instrument that helps you multiply. (3,9)

8. This is a body with only bones. (2, 2)

9. This is a large head from Greek Mythology. (1,7)

Answer Key

Page 13—Animal Life Span

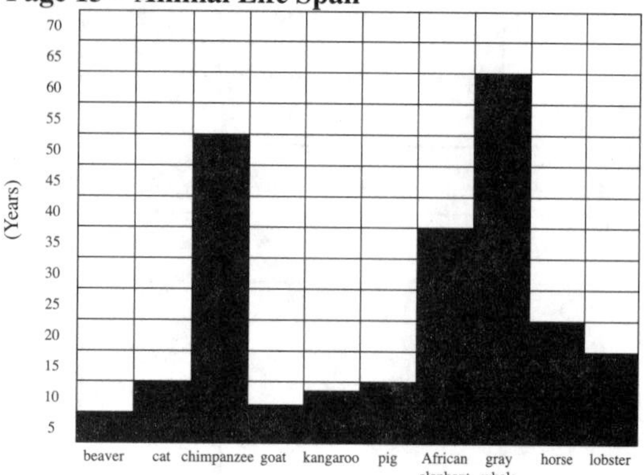

1. gray whale
2. beaver
3. cat and lobster
4. 63 years longer
5. beaver
6. alligator or humpback whale

Page 15—Biodiversity in Species

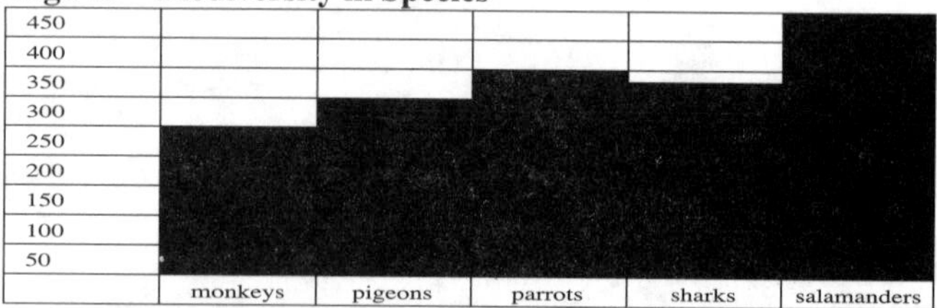

1. salamanders
2. monkeys
3. 100
4. 200
5. 100

Page 17—Tallest Buildings in the World

120
110
100
90
80
70
600
50
Citic Plaza
Jin Maocc Tower
Petrons Towers
Sears Tower
Two International Finance Center
Taipei 101

1. Sears
2. Jin Mao, Petrons, and Two International Finance
3. 20
4. none
5. 270

Page 19—Disasters

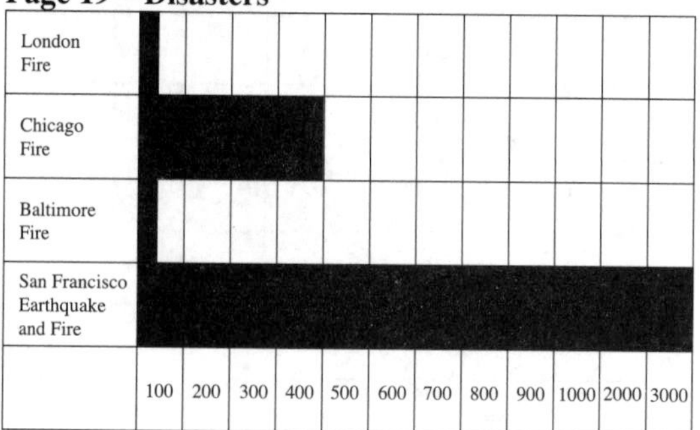

1. San Francisco
2. Baltimore and London
3. 296
4. 2,700
5. 3,308

Page 21—Smallest Population per Square Mile

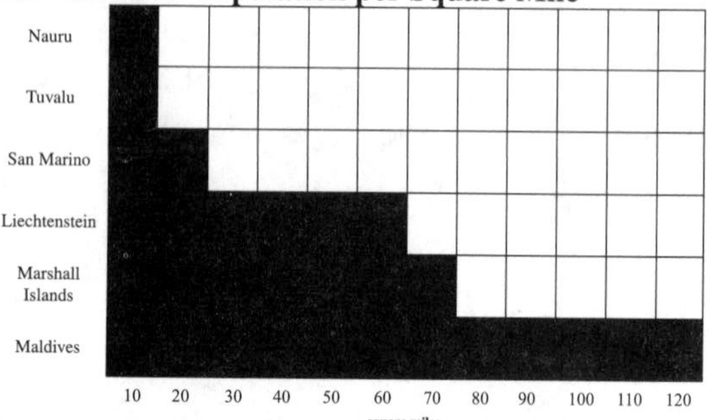

1. Nauru and Tuvalu
2. Maldives
3. Nauru, Tuvalu, San Marino
4. 50
5. Maldives

Page 23—Number of Letters in the States

4	5	6	7	8	9	10	11	12	13
1/25	2/25	3/25	5/25	8/25	2/25	1/25	2/25	0/25	1/25

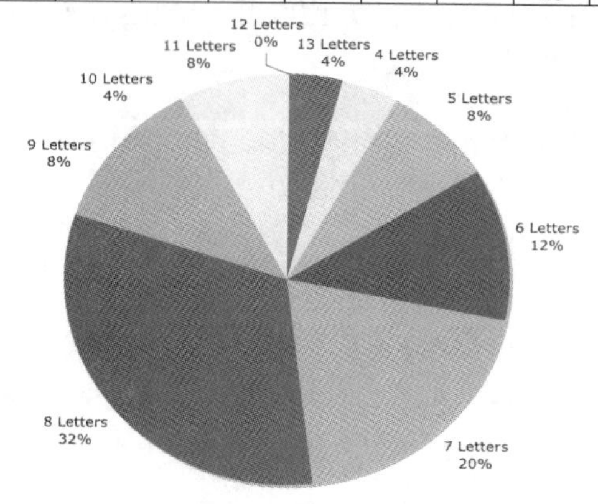

1. 8%
2. 0%
3. 20%
4. 20%
5. 52%

Answer Key *(cont.)*

Page 25—Top Energy Users

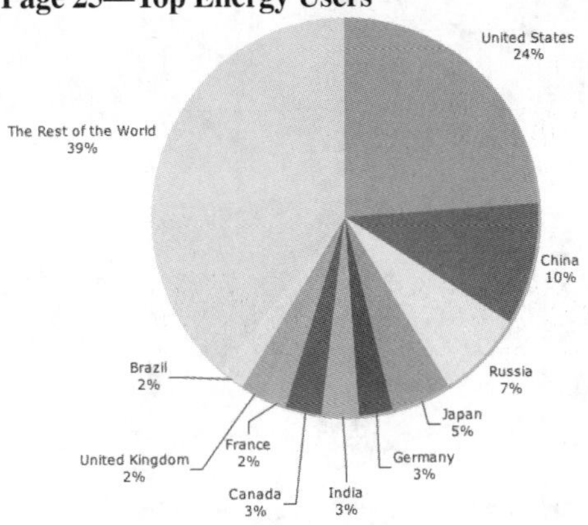

1. United States
2. France, United Kingdom and Brazil
3. 17%
4. 21%
5. Answers will vary, might include: We are a large country, have many industries, are technologically advanced, extremely wealthy, often wasteful with our resources

Page 27—Where Garbage Goes

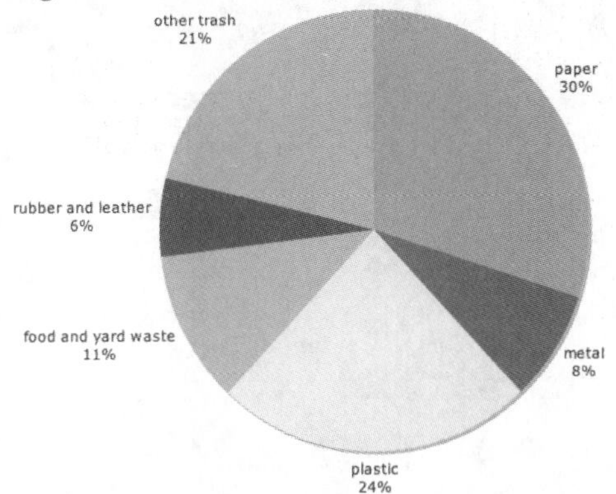

1. 32%
2. 30%
3. plastic
4. 21%
5. answers will vary

Page 29—Growing U.S. Population

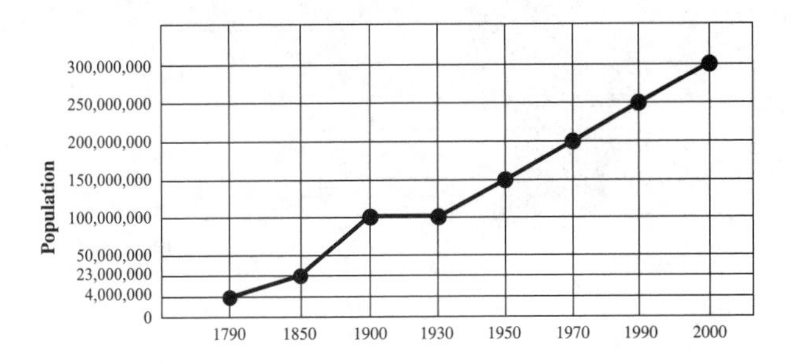

1. 5,000,000
2. 15,000,000
3. 395,000,000
4. 50,000
5. immigrants/immigration

Page 31—Grand Slams

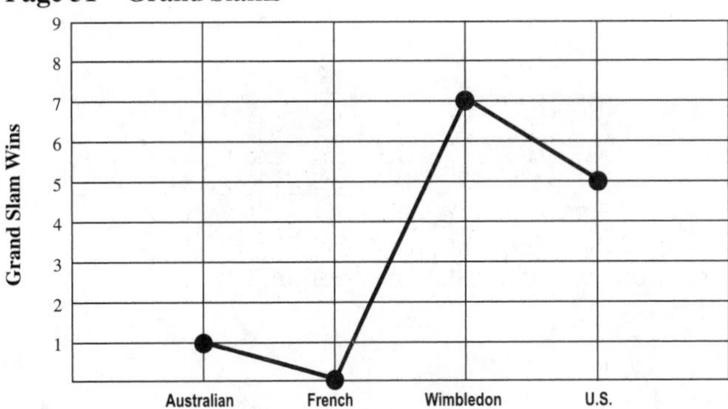

1. 2
2. 7
3. 7
4. 5
5. 11

Page 33—Symbols of the United States

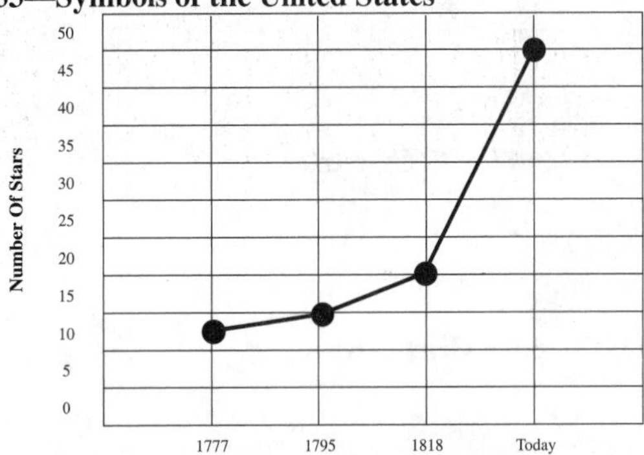

1. 50
2. 1
3. 30 stars
4. 5 stars
5. 2

Page 35—How Fast Do Animals Run?

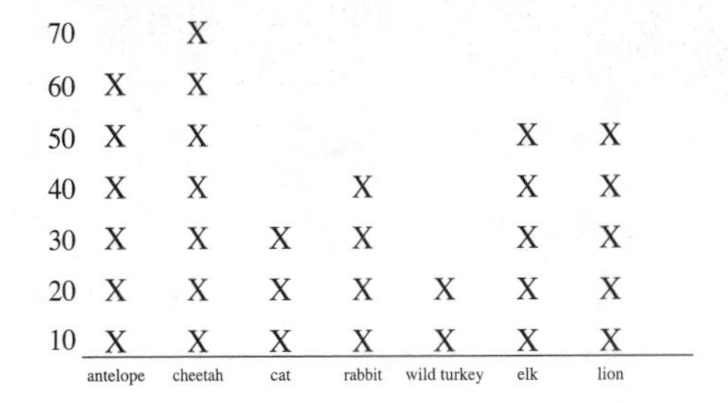

1. 70 miles per hour
2. 10 miles per hour
3. cheetah, antelope
4. wild turkey
5. 20 miles per hour

Answer Key *(cont.)*

Page 37—Kinds of Musical Instruments

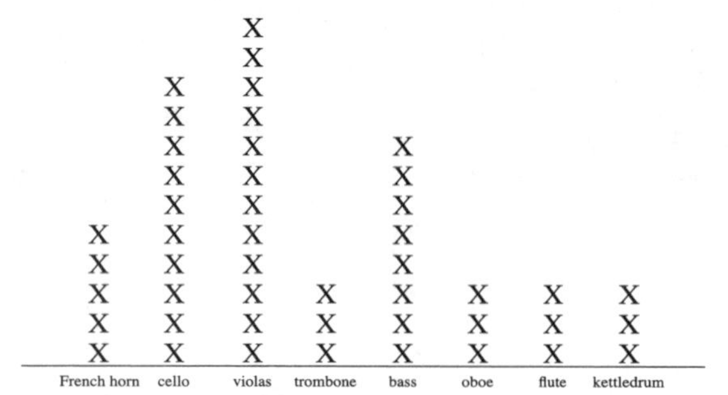

1. viola
2. trombone, oboe, flute, kettledrum
3. 7
4. 5
5. 17

Page 39—Record Temperatures

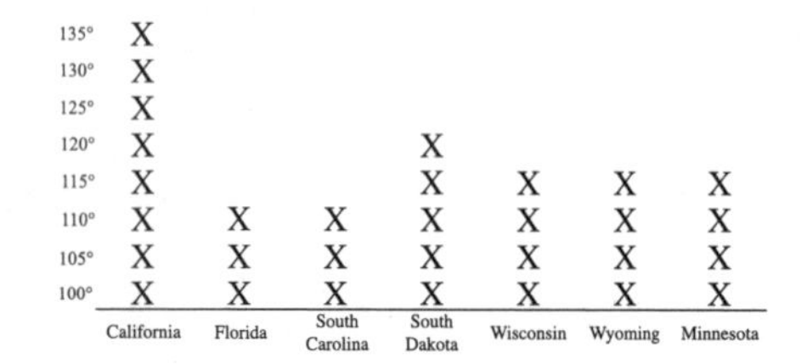

1. California
2. Florida and South Carolina
3. 20 degrees
4. 10 degrees
5. The temp is 5 degrees lower.

Page 41—Dinosaur Celebrities

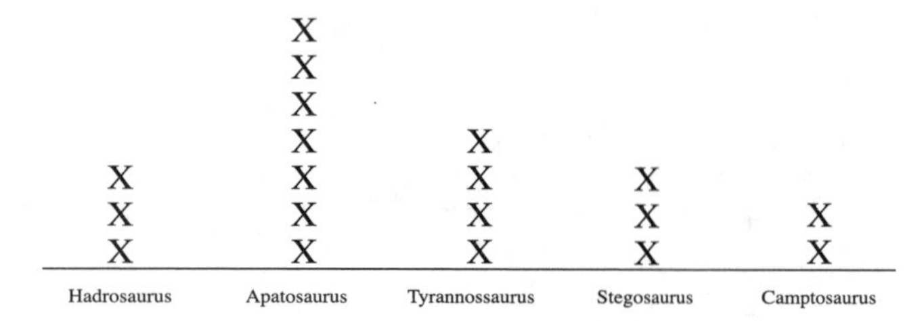

1. camptosaurus
2. apatosaurus
3. 10 feet
4. 10 feet
5. camptosaurus

Page 43—Largest Cities in the World

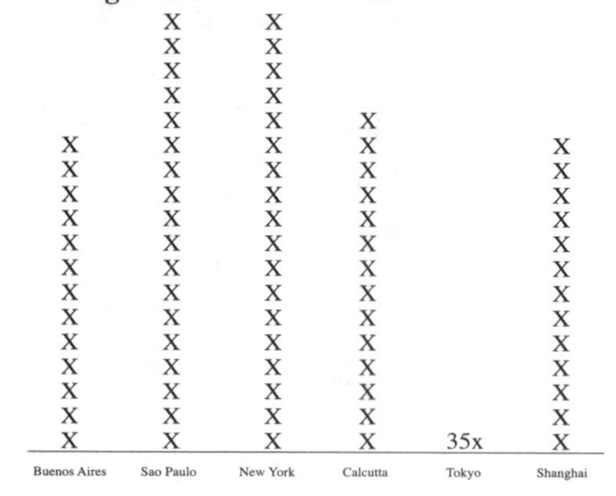

1. Tokyo
2. Shanghai and Buenos Aires
3. 5 million
4. 4 million
5. 62 million

Page 45—Tennis Tournaments

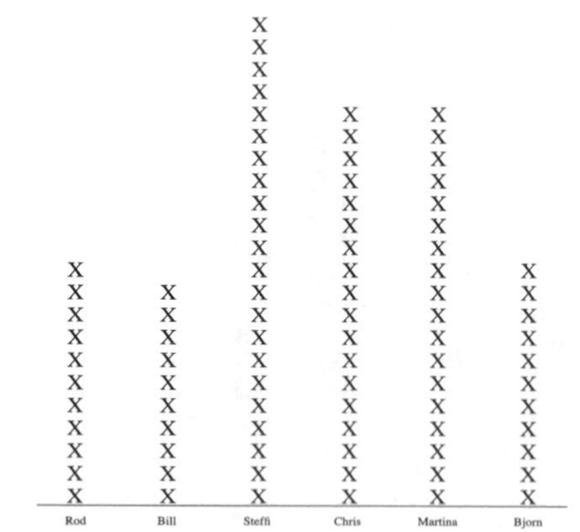

1. Steffi Graf
2. Bill Tilden
3. 12
4. 90
5. Women. The men have only 32 and the women have 59.

Page 47—Exercise—It's What You Do!

Bicycling (9.4 mph)	☺☺☺☺☺
Bicycling (5.5 mph)	☺☺☺
Jogging (6 mph)	☺☺☺☺☺☺☺☺☺
Jumping rope (easy pace)	☺☺☺☺☺☺☺☺
Playing basketball	☺☺☺☺☺☺☺☺
Playing soccer	☺☺☺☺☺☺☺
Rollerblading	☺☺☺☺
Skiing (downhill)	☺☺☺☺☺
Swimming (25 ypm)	☺☺☺
Walking (3mph)	☺☺☺

1. Bicycling (5.5 mph), Swimming, and Walking
2. 9.4
3. Rollerblading
4. Jumping rope and playing basketball
5. One more calorie per minute

Answer Key *(cont.)*

Page 49—Facts About Nations

Tunisia	[icon] [icon]
Turkey	[icon] [icon] [icon]
Turkmenistan	[icon] [icon] [icon]
Tuvalu	[icon] [icon]
Uganda	[icon] [icon] [icon]
Ukraine	[icon] [icon]
United Arab Emirates	[icon] [icon] [icon] [icon]
Great Britain	[icon]

1. United Arab Emirates
2. Great Britain (United Kingdom)
3. 5
4. 2
5. 1

Page 51—November Birthdays

Authors	X
Music Careers	XXXXXXX
TV Hosts/Actors	XXXXXXXXXX
Sports Careers	XXX
Inventor/Scientist	XX
Outlaw	X
Government/Politics	X

1. Ty Host/Acting
2. Government/Politics
3. 7
4. 10
5. Billy the Kid

Page 53—Golf

Patty Berg	XXXXXXXXXXXXXXX
Walter Hagen	XXXXXXXXXXX
Ben Hogand	XXXXXXXXX
Jack Nicklaus	XXXXXXXXXXXXXXXXXX
Gary Player	XXXXXXXXX
Betsy Rawls	XXXXXXXX
Annika Sorenson	XXXXXXXX
Louise Suggs	XXXXXXXXXXX
Tom Watson	XXXXXXXX
Tiger Woods	XXXXXXXXX
Mickey Wright	XXXXXXXXXXXXX

1. Patty Berg
2. 3
3. Annika Sorenson and Betsy Rawls
4. 1
5. 65

Page 55—Major Cultural Areas of the Native North Americans

Arctic	X X
California	X X X X X X X X
Great Basin	X X X
Northeast Woodlands	X X X X X X X X
Northwest Coast	X X X X X X X X X
Plains and Prairie	X X X X X X X
Plateau	X X X X X X X X
Southeast Woodlands	X X X X X
Southwest	X X X X X X
Subarctic	X X X X X X X X X X

1. Subarctic
2. Arctic
3. 14
4. 1
5. 66

Page 57—Body Basics

respiratory	X X X
endocrine	X X X
digestive	X X X X
circulatory	X X X X
nervous	X X X X X X

1. nervous
2. endocrine and respiratory
3. 3
4. 9
5. 20

Page 59—The Planets

Earth	Juniper	Mars	Mercury	Neptune	Pluto	Saturn	Uranus	Venus
1	63	2	0	13	1	34	27	0

0 , 0 , 1 , 1 , 2 , 13 , 27 , 34 , 63

Moons around the Planets

Stem	Leaves
0	0 1 2 3
1	3
2	7
4	7
6	3

Key: 1 3 = 13 moons on planets

1. 0
2. 63
3. clustered
4. Answers will vary.

Page 61—Comparing Money

Luxembourg	Moldova	Mauritius	Mexico	Nepal	Nicaragua	Kenya	Kyrgyzstan
77	13	29	11	71	16	77	41

11,13, 16, 29, 41, 71, 77, 77

Currency

Stem	Leaves
1	1 3 6
2	9
4	1
7	1 7 7

Key: 2 9 = 29 currency

1. Mexico
2. Luxembourg
3. clustered
4. Answers will vary.

Answer Key *(cont.)*

Page 63—Basketball

Alan Iverson	Andrei Kirilenko	Kevin Garnett	Larry Hughes	Steve Nash
75	41	82	61	75

41 , 61 , 75 , 75 , 82

Games Played

Stem	Leaves
4	1
6	1
7	5 5
8	2

Key: 4 1 = 41 games played

1. 41
2. 82
3. clustered
4. Answers will vary.

Page 65—Unique Holidays

Pooh Day	Ferris Wheel Day	Pecan Day	Sibling Day	Save the Rhino Day	Handshake Day
Jan. 18	Feb.14	March 25	April 10	May 1	June 28

Ice Cream Day	Sandcastle Day	Talk like a Pirate Day	National Knock Knock Day	World Hello Day	Make Up Your Mind Day
July 17	Aug. 17	Sept.19	Oct. 31	Nov. 21	Dec. 31

1. Feb. 14
2. June 28
3. the beach
4. It is National Ice Cream Day.
5. Knock- knock jokes

Page 67—Famous Scientists

Archimedes	Nicolas Copernicus	Sir Isaac Newton	Charles Darwin	George Washington Carver
287 BC	1473	1642	1809	1864

Albert Einstein	Rachel Carson	Jane Goodaell	Stephen Hawking	Linda Spilker	Paul Sereno
1879	1907	1934	1942	1955	1957

1. Archimedes
2. Linda Spilker and Paul Sereno
3. Charles Darwin, George Washington Carver, and Albert Einstein
4. Only one, Archimedes
5. Rachel Carson, Jane Goodall, Stephen Hawking, Linda Spilker, and Paul Sereno

Page 69—Cycling. . . Tour de Lance

1. 1971
2. 1996
3. 1985
4. US Postal Service (racing) Team
5. 33

Page 71—Kid Inventors

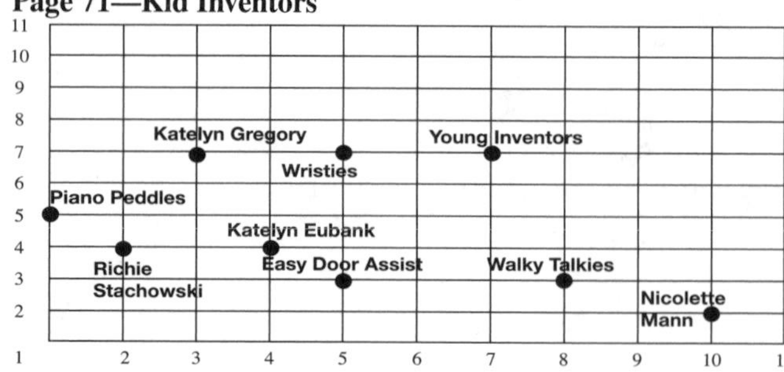

1. Kathryn Gregory (3,7)
2. Richie Stachowski (2,4)
3. Nicolette Mann (10,1)
4. Katelyn Eubank (4,3)
6. Wristies (5,7)

Page 73—Getting to the Root

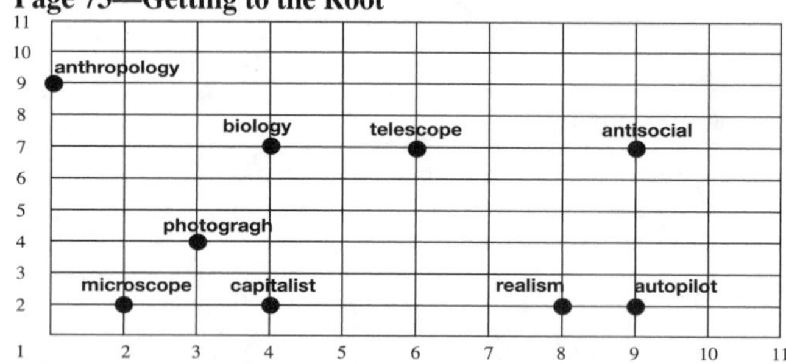

1. capitalist (4,2)
2. autopilot (9,2)
3. photograph (3,4)
4. biology (4,7)

Page 75—Joke and Riddles

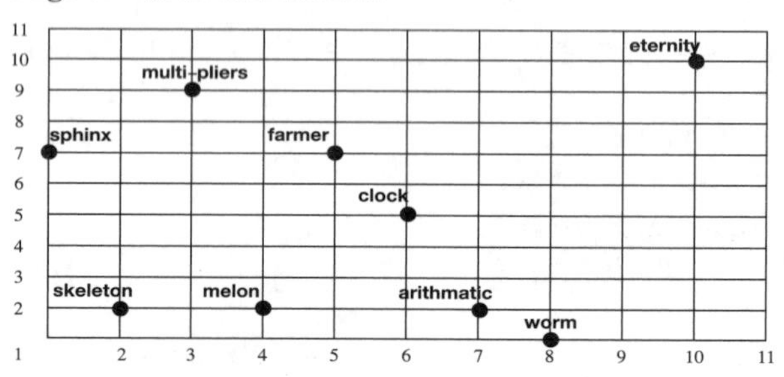

1. clock (6,5)
2. arithmetic (7,2)
3. worm (8,1)
4. eternity (10,10)
5. farmer (5,7)

